# CHARACTERIZATION OF EPITAXIAL SEMICONDUCTOR FILMS

METHODS AND PHENOMENA
THEIR APPLICATIONS IN SCIENCE AND TECHNOLOGY



Volume 1. A.W. Czanderna (Ed.), Methods of Surface Analysis
Volume 2. H. Kressel (Ed.), Characterization of Epitaxial Semiconductor Films

*In preparation*

W.R. Bottoms (Ed.), Scanning Electron Microscopy

F.J. Fry (Ed.), Ultrasound: Its Application in Medicine and Biology, Parts I and II

# CHARACTERIZATION OF EPITAXIAL SEMICONDUCTOR FILMS

Edited by

HENRY KRESSEL

*RCA Laboratories,*
*Princeton, N.J. 08540, U.S.A.*

VOLUME 2 of

Methods and Phenomena: Their Applications in Science and Technology

This set of papers has been published as a special
issue of *Thin Solid Films*, Vol. 31, issues 1 and 2

ELSEVIER SCIENTIFIC PUBLISHING COMPANY
Amsterdam — Oxford — New York 1976

ELSEVIER SCIENTIFIC PUBLISHING COMPANY
335 Jan van Galenstraat
P.O. Box 211, Amsterdam, The Netherlands

AMERICAN ELSEVIER PUBLISHING COMPANY, INC.
52 Vanderbilt Avenue
New York, New York 10017

ISBN: 0-444-41438-x

With 128 illustrations and 15 tables



Printed in The Netherlands

## CONTRIBUTORS TO VOLUME 2


| | |
|---|---|
| G.R. Booker | Department of Metallurgy, University of Oxford, Oxford, Gt. Britain |
| A.G. Cullis | Bell Laboratories, Murray Hill, N.J., U.S.A. |
| R.E. Honig | RCA Laboratories, Princeton, N.J., U.S.A. |
| S.H. McFarlane III | RCA Laboratories, Princeton, N.J., U.S.A. |
| D.C. Miller | Bell Laboratories, Murray Hill, N.J., U.S.A. |
| W.V. Muench | Institut für Werkstoffkunde, Technische Universität, Hannover, Germany |
| I. Pfaffeneder | Institut für Werkstoffkunde, Technische Universität, Hannover, Germany |
| G.A. Rozgonyi | Bell Laboratories, Murray Hill, N.J., U.S.A. |
| K.V. Ravi | Mobil Tyco Solar Energy Corporation, Waltham, Mass., U.S.A. |
| G.E. Stillman | Lincoln Laboratory, Massachusetts Institute of Technology, Lexington, Mass., U.S.A. |
| D.J. Stirland | The Plessey Company, Allen Clark Research Centre, Towcester, Northants., Gt. Britain |
| B.W. Straughan | Royal Radar Establishment, Great Malvern, Worcs., Gt. Britain |
| C.C. Wang | RCA Laboratories, Princeton, N.J., U.S.A. |
| H.H. Wieder | Naval Electronics Laboratory Center, San Diego, Calif., U.S.A. |
| C.M. Wolfe | Lincoln Laboratory, Massachusetts Institute of Technology, Lexington, Mass., U.S.A. |
| J.J. Young | Moore School of Electrical Engineering, University of Pennsylvania, Philadelphia, Pa., U.S.A. |
| J.N. Zemel | Moore School of Electrical Engineering, University of Pennsylvania, Philadelphia, Pa., U.S.A. |

METHODS AND PHENOMENA

# PREFACE

These invited papers are concerned with important aspects of semiconductor epitaxial film characterization by electron microscopy and electrical, chemical and X-ray techniques. The semiconductors discussed include silicon, SiC, PbSe and various III—V compounds, but the techniques are in most cases of broad utility for other materials.

Thin semiconductor films epitaxially deposited on large-area single-crystal substrates constitute the fundamental building blocks of modern electronic and optoelectronic devices which include integrated circuits, power and microwave transistors, light emitting diodes and radiation detectors. The detailed materials requirements differ vastly, of course, both with regard to the type of semiconductor needed, the degree of material perfection and the impurity concentration. Independent of the material, however, the determination of relevant electrical and structural parameters is essential for the successful utilization of the epitaxial layers for device fabrication. In the past decade, much progress has been made in all aspects of the characterization techniques for thin films. Electron microscopy, X-ray topography and impurity analysis have received particular emphasis. Furthermore, added sophistication in traditional electrical measurements and their interpretation has yielded more accurate information than previously possible. Etching techniques to reveal defects have been refined for important materials such as GaAs, and a wealth of information can now be revealed by these simple methods. The determination of native point defect concentrations, however, is still very difficult, but novel methods involving gas—solid interactions may provide a useful tool for their study.

I wish to acknowledge the Editors of *Thin Solid Films*, and in particular Professor J. Zemel, for their support in the preparation of this special issue, now published in book form.

RCA Laboratories
Princeton, N J.

Henry Kressel

# CONTENTS

*Thin Solid Films*, *31* (1976) 3–23



# CRYSTAL GROWTH AND DEFECT CHARACTERIZATION OF HETEROEPITAXIAL III–V SEMICONDUCTOR FILMS

C. C. WANG AND S. H. MCFARLANE III
*RCA Laboratories, David Sarnoff Research Center, Princeton, New Jersey (U.S.A.)*
(Received January 6, 1975; accepted June 17, 1975)

The growth of heteroepitaxial III–V compounds on dielectric oxide substrates has been most successfully and reproducibly achieved by the organometallic growth process developed in recent years. The epitaxial film growth characteristics are largely affected by several key parameters including the substrate orientation, substrate surface preparation, growth temperature, gas flow and source material purity. The heteroepitaxial films have been characterized by X-ray diffraction topography and found to be composed of characteristic grain structures which are misoriented $\pm 0.1°$ from the nominal orientation of the layer. The film/substrate composites are deformed, and the films are generally under compressive stress of the order of $10^9$ dyn $cm^{-2}$. Electrical properties of the heteroepitaxial films are inferior to the bulk material properties, but are favorable enough for applications in several devices with potential advantages.

## 1. INTRODUCTION

The achievement of single-crystal growth of large area semiconductor films on substrates of different materials is of technical importance to many solid state electronic devices. The heteroepitaxial composite structure is also of scientific interest because the epitaxy is determined by the spatial relationship between the atomic arrangement in the substrate and that of the atoms in the appropriate crystallographic plane of the semiconductor. The degree of crystalline perfection of the semiconductor deposits also depends largely on the physical/chemical nature of the substrate surfaces.

Extensive studies of the epitaxy of elemental semiconductors (silicon and germanium) on sapphire ($\alpha$-$Al_2O_3$) and spinel ($MgAl_2O_4$) oxide substrates have been reported[1–4] in the past few years. These have led to a basic understanding of the composite material systems and to the development of the silicon-on-sapphire technology. Currently, the trend of development has extended to the III–V compound semiconductors. However, because of the complexity in material growth and device fabrication, development of III–V film/oxide substrate composite systems is still at an early stage. Research studies on epitaxial growth by new techniques, such as the organometallic method[5, 6], and on defect characterization of the composite materials have been increasing in many leading laboratories. Several device structures have been successfully fabricated in the heteroepitaxial systems with interesting results. These include GaAs/sapphire transmission mode photocathodes[7], GaN/sapphire, GaAs/spinel and GaP/spinel electroluminescent diodes[8–11], GaAs/sapphire Gunn oscillators[12], GaAs/BeO microwave transistors[13] and AlN/sapphire surface acoustic wave devices[14].

In this paper, a state-of-the-art report is presented to review the recent and current developments on the epitaxial growth and defect characterization of heteroepitaxial III–V semiconductor films grown on dielectric oxide substrates. The main topics to be discussed in the following sections include the effect of growth parameters on epitaxial film perfection, the characterization of defects by X-ray diffraction topography and the electrical properties of the films. Emphasis is placed on the characteristics of GaAs and GaP films, which have received the most study to date.

## 2. EPITAXIAL GROWTH PROCESS

The epitaxial growth of III–V compounds on oxide substrates has been most successfully and reproducibly achieved using the organometallic vapor phase growth (OMV) process[5, 6]. Single-crystal GaAs films grown on several oxide substrates by the reaction between trimethyl gallium ($(CH_3)_3Ga$) and arsine ($AsH_3$) were first reported[15] in 1968. Since then the OMV process has received increasing attention, and the epitaxy of many III–V compounds (nitrides, phosphides, arsenides and antimonides) and their solid solutions on dielectric oxide substrates has been achieved. Attempts using the conventional chloride vapor phase epitaxial growth (VPE) process and the liquid phase epitaxial growth (LPE) process have been less successful and/or not reproducible. A survey of the literature revealed that, among the III–V compounds, only the hexagonal nitrides (AlN [16] and GaN [17]) were reported to form successful epitaxial films on sapphire substrates using the VPE process, and that (Ga, Al)As was reported[18] to be epitaxially grown on $MgGa_2O_4$ by the LPE method. For the cubic III–V compounds, such as GaAs, the growth by the VPE process of a continuous layer less than 2 μm thick on sapphire or spinel was reported[19] to be extremely difficult, and the epitaxial yield was only about 30%.

The difficulty encountered in the epitaxial growth of III–V compounds on oxides by the conventional processes is primarily due to the sensitivity of the substrate to the chemical nature of the growth systems. Experimental results have shown that the OMV process is least sensitive to the condition of the substrate surface, and is therefore most successful in achieving heteroepitaxy. To date, the OMV process has been used almost exclusively to grow III–V epitaxial films on oxide substrates. In this process, III–V compounds and alloys are produced by decomposing appropriate group III organometallic compounds (generally group III alkyls) in the presence of the appropriate group V hydrides or alkyls. In addition, the group II organometallics and group VI hydrides are commonly used as dopants for the III–V compounds. The reaction leading to the growth of GaAs, for example, is

$$(CH_3)_3Ga + AsH_3 = GaAs + 3CH_4$$

$H_2S$ (or $H_2Se$) and $(C_2H_5)_2Zn$ can be conveniently used as dopants for n and p epitaxial GaAs films, respectively.

In addition to the ability of achieving epitaxial growth of compound films readily on foreign substrates, there are also other distinct advantages of the organometallic process. The process requires only one controlled hot temperature zone for the *in situ* formation and growth of the compound semiconductors on heated substrates and occurs in an atmosphere free of halide-containing species which may undesirably etch certain film/substrate systems. The epitaxial growth can be achieved at relatively low temperatures (*e.g.* 650 °C for GaAs) and alloy films with various compositions can be conveniently grown with minimum additional growth facilities. However, there are also drawbacks. The control of impurities in the organometallic source materials is difficult, and therefore at low carrier density ($< 10^{16}$ cm$^{-3}$) the control of doping is not easy to achieve. In certain cases this may be complemented by the conventional growth processes. For example, two stage epitaxial growth has been achieved in growing GaAs and GaP[9–11] electroluminescent diodes on oxide substrates. This involves the initial growth by the OMV process of the III–V epitaxial layer on the substrate, followed by the LPE of appropriately doped layers to form the p–n junctions. This approach is made because of the ease of initiating growth inherent in the OMV process and because of the advantages found in the LPE methods, including the control of stoichiometry, gettering of impurities, convenient incorporation of dopants and possible improvement of the crystalline perfection of the III–V overgrowth.

## 3. EFFECT OF GROWTH PARAMETERS ON EPITAXIAL FILM CHARACTERISTICS

The important growth parameters that affect the heteroepitaxial film characteristics are: substrate orientation, substrate crystallinity and surface preparation, growth temperature, gas flows, reactor geometry, source material quality and growth procedure. The effect of the various parameters on the layer properties depends on the growth process and may vary from one material system to another. The discussions which follow place major emphasis on the growth of GaAs on various oxides by the organometallic process to illustrate the effect of the growth parameters.

### 3.1. *Substrate orientation*

In heteroepitaxial systems, the substrate orientation determines the film/substrate orientation relationships, and it also plays a major role in the ease with which the epitaxial films grow. In the GaAs/spinel system, several film/substrate orientation relationships have been determined[20, 21]. The epitaxy is unusual in that from the crystal structure point of view (both GaAs and $MgAl_2O_4$ are cubic and belong to the same space group) one would expect a unique parallel epitaxial relationship to exist between the film and substrate, as in the case of Si on spinel[22]. However, the experimentally determined relationships are the isostructural (111)GaAs||(111)$MgAl_2O_4$ and (100)GaAs||(100)$MgAl_2O_4$, and the unexpected (100)GaAs||(110)$MgAl_2O_4$. Also, an apparent (111)GaAs||(100)$MgAl_2O_4$ relationship exists when the (100) $MgAl_2O_4$ substrate is off-oriented by 1° or more.

Concurrently, it was found that the ease of growing epitaxial GaAs on spinel depends critically on the substrate orientation. This is again in contrast to the Si-on-spinel growth case in which simultaneous epitaxial growth of Si can be achieved on spinel substrates of different orientations. The ease of GaAs growth increases with spinel orientation in the order (100)<(110)<(111). The epitaxy of (100) GaAs on (110) spinel is more sensitive to substrate surface quality and to growth temperature than that of (111) GaAs on (111) spinel. The growth on (100) spinel is most difficult in that polycrystalline films (with isolated (100) GaAs single-crystal areas) are generally formed in the various growth conditions studied. However, with a substrate off-orientation as small as 1° the growth is enhanced remarkably, with smooth single-crystal GaAs films formed on the off-oriented (100) spinel. The GaAs exhibits an unexpected (111) orientation. The ease of growth of (111) GaAs on off-oriented (100) spinel is comparable with that of (111) GaAs on (111) spinel. For the (111) and (110) substrates, (111) and (100) films, respectively, with improved surface crystallinity have been grown on slightly off-oriented (1°–3°) substrate surfaces. Stereographic projections of (100) GaAs on (110) spinel and (111) GaAs on (100) spinel are presented in Figs. 1 and 2, showing the unusual orientation relationships. Surface structures of GaAs films grown on (111), (110), (100) and off-oriented (100) spinel substrates are shown in Fig. 3.

The effect of substrate orientation on the film growth characteristics was also studied in the GaP/spinel system[11]. It is similar to the GaAs/spinel system in that the ease of growth increases with spinel orientation in the order (100)<(110)<(111). However, it differs from the GaAs/spinel system in that the GaP epitaxial films

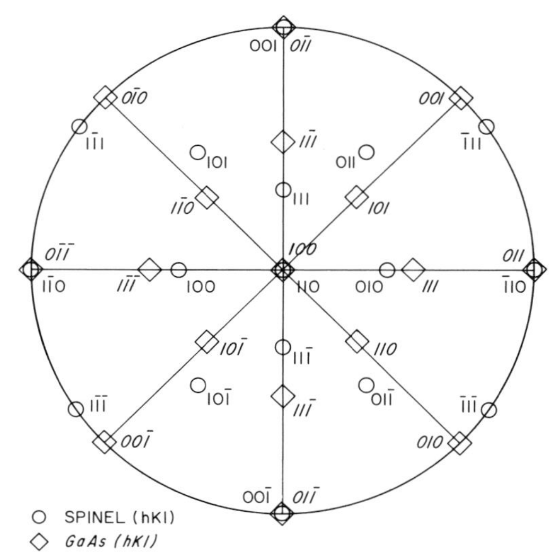


Fig. 1. Stereographic projection of (100) GaAs on (110) spinel.

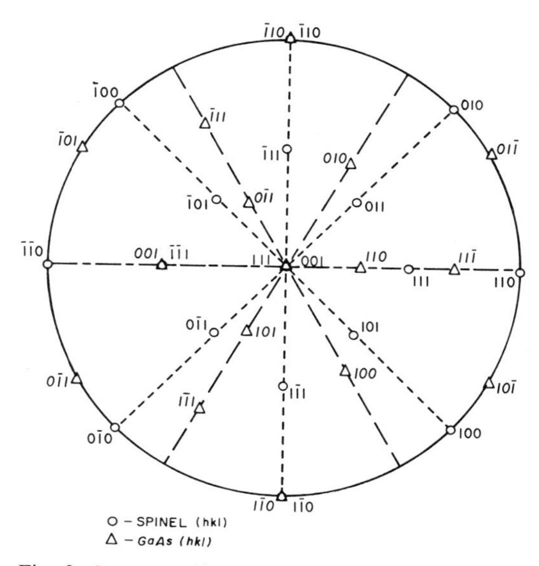


Fig. 2. Stereographic projection of (111) GaAs on (100) spinel.

grown on (110) spinel substrates are composed of both (110) and (100) grains, instead of the (100) orientation only as in the GaAs/spinel epitaxy. The results indicate that two orientations of the GaP overgrowth fulfil the condition for a single substrate orientation. Therefore the GaP film is not a single crystal, but is still epitaxial. Similar situations (epitaxial but not single-crystal growth) are encountered in many other heteroepitaxial film/substrate systems. The GaP/ spinel system also differs from the GaAs/spinel system in that the GaP films exhibit no significant difference in growth characteristics (surface orientation and/or crystallinity) as a function of the degree of substrate off-orientation.

The strong dependence of GaAs film crystalline quality on substrate orientation has also been reported in the GaAs/sapphire and GaP/sapphire systems[20, 23]. The (0001) sapphire basal plane is least sensitive to surface preparation and to growth conditions. The physical and electrical characteristics of (111) GaAs films grown on (0001) sapphire have been studied[24] extensively. In contrast to the (0001) orientation, it is relatively difficult to grow single-crystal GaAs and GaP of good quality on other sapphire orientations. In the GaAs/BeO system, among nine substrate orientations studied it was found[13] that the (10$\bar{1}$1) and (11$\bar{2}$2) BeO surfaces tended to yield better quality (111) and (100) GaAs films, respectively.

Table I gives an up-to-date summary of the epitaxial relationships of III–V compound/oxide substrate composite systems with substrate orientations that are known to yield films of good quality. In spite of the unusual orientation relationships found in some systems, there seem to be a few generalized trends observed from the compound semiconductor/oxide heteroepitaxial relationships. First, for films with cubic crystal structures, the relationships (111)film||(0001)-sapphire and (111)film||(111)spinel have generally been found. In fact, for most

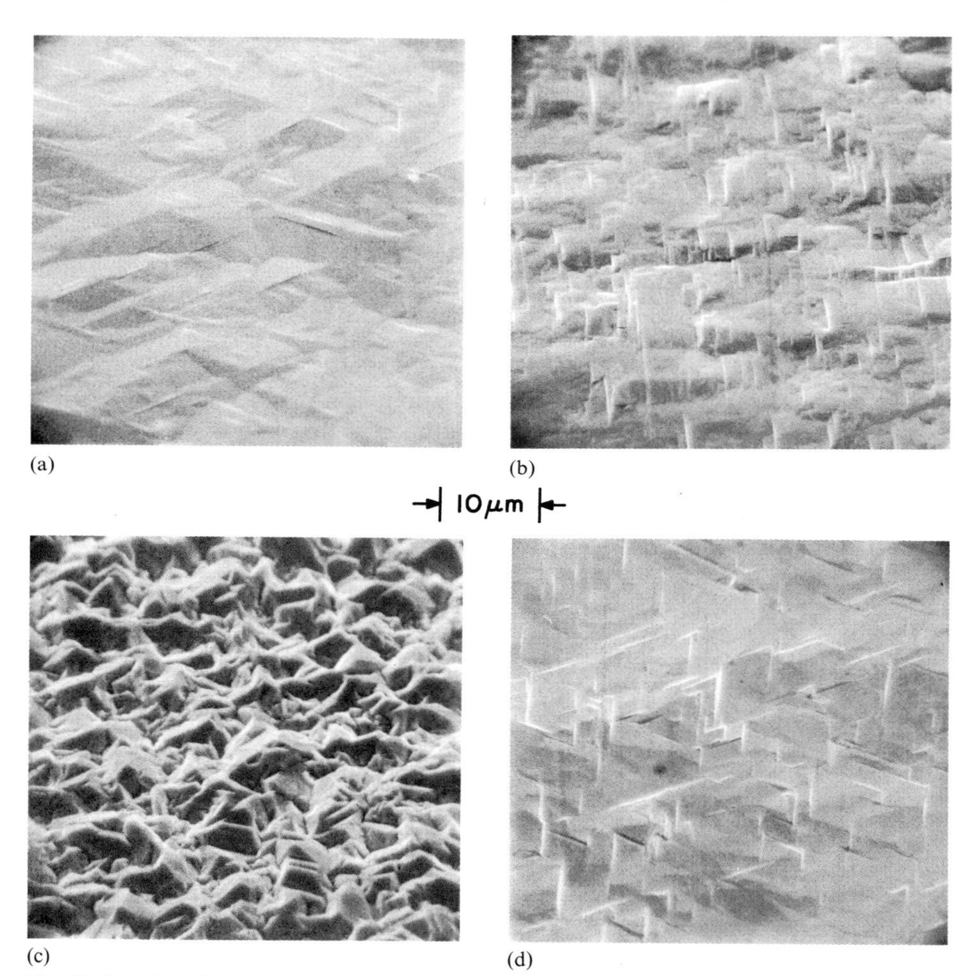


Fig. 3. Scanning electron micrographs of GaAs films (~20 μm), beam–sample angle 45°: (a) on a (111) spinel substrate; (b) on a (110) spinel substrate; (c) on a (100) spinel substrate; (d) on a 1° off-oriented (100) spinel substrate.

film materials the (111) epitaxial films can be grown more easily than any other orientations under a wide range of growth conditions. Second, the (0001) sapphire orientation (the basal plane) has yielded the most cases of successful epitaxy for the compound semiconductor films investigated to date. In the cubic III–V compound/sapphire systems, such as GaAs-on-sapphire, the (0001) surface is the only orientation that yields films of good quality. The $(01\bar{1}2)$ orientation (the cleavage plane) most commonly used in Si-on-sapphire systems (yielding (100) Si) has not been used successfully in III–V/sapphire epitaxy. Films grown on this orientation exhibit (111) orientation and are of poor quality. In the III nitride composites, *e.g.* the AlN-on-sapphire system[25, 26], both (0001) and $(01\bar{1}2)$ can be used to yield films of good quality. The $(01\bar{1}2)$ orientation is preferred

TABLE I

FILM/SUBSTRATE ORIENTATION RELATIONSHIPS (III–V COMPOUNDS ON OXIDE SUBSTRATE PLANES WHICH YIELD HIGH QUALITY FILMS)

| *Film* | *Substrate* | *Parallel orientations film/substrate* | *Reference* | *Remarks* |
|---|---|---|---|---|
| AlN | $\alpha$-$Al_2O_3$ | (0001)//(0001); $[\bar{1}210]//[\bar{1}100]$ | 25,26 | – |
| AlN | $\alpha$-$Al_2O_3$ | $(11\bar{2}0)//(01\bar{1}2)$; $[0001]//[01\bar{1}1]$ | 25,26 | Films used for SAW devices[14] |
| GaN | $\alpha$-$Al_2O_3$ | (0001)//(0001); $[\bar{1}210]//[\bar{1}100]$ | 25,26 | – |
| GaN | $\alpha$-$Al_2O_3$ | $(11\bar{2}0)//(01\bar{1}2)$; $[0001]//[01\bar{1}1]$ | 25,26 | Films used for electroluminescent diodes[8] |
| GaP | $\alpha$-$Al_2O_3$ | (111)* //(0001); $[1\bar{1}0]//[11\bar{2}0]$ | 23 | – |
| GaP | $MgAl_2O_4$ | (111)//(111); $[01\bar{1}]//[01\bar{1}]$ | 11,23 | Films used for electroluminescent diodes on a two-stage epitaxy[11] |
| GaP | Si/$\alpha$-$Al_2O_3$ | (100)//(100)//; $(01\bar{1}2)$ | 23 | Films potentially useful as photocathode substrates |
| GaAs | $\alpha$-$Al_2O_3$ | (111)//(0001); $[1\bar{1}0]//[11\bar{2}0]$ | 20 | Films used for microwave diodes[12] |
| GaAs | $MgAl_2O_4$ | (111)//(111); $[01\bar{1}]//[01\bar{1}]$ | 20,21 | Films used for electroluminescent diodes on a two-stage epitaxy[9] |
| GaAs | $MgAl_2O_3$ | (100)//(110); $[011]//[\bar{1}10]$ | 20,21 | Films can be grown with same quality as (111) GaAs[21] |
| GaAs | $MgAl_2O_4$ | (111)//(100); $[01\bar{1}]//[\bar{1}10]$ | 21 | Films exhibit same quality as (111) films on (111) spinel[21] |
| GaAs | BeO | $(100)//(11\bar{2}2)$; – | 13 | Films used for microwave transistors[13] |
| GaAs | BeO | $(111)//(10\bar{1}1)$; – | 13 | – |

* All the (111) heteroepitaxial III–V films studied exhibit the (111) A face.

for AlN-on-sapphire SAW devices[14]. The (0001) sapphire has been used most for many film materials to date and conceivably will be employed for the study of film growth of new composite systems.

It is interesting to note further that the dependence on substrate orientation of the ease of growth of heteroepitaxial semiconductors on oxide substrates is much less critical for the elemental semiconductors (such as Si) than for the compound semiconductors (such as GaAs). For example, single-crystal silicon films can conveniently be grown simultaneously on the major cubic spinel surfaces or on several common sapphire orientations. Another interesting aspect is that some film/substrate epitaxial orientation relationships are different between the Si/oxide and GaAs/oxide systems despite the similar crystal structure and lattice parameters for Si and GaAs. This suggests that the film/substrate interfacial bondings between the Si/oxide and the GaAs/oxide composite systems are quite different. In fact, it is more than likely that the nature of the interfacial bonding in a given compound semiconductor/oxide composite system may vary from one mechanism to another depending on the film/substrate orientation.

The film/substrate orientation relationships have practical significance for considerations in device feasibility. From the crystallinity point of view, a substrate orientation should be chosen so that the best film epitaxial growth may be achieved. On the other hand, there are devices for which only a certain orientation is desirable. Therefore films with the best quality may not exhibit the most desirable orientation for devices. In many cases, an alternative approach may be made to fulfil both the crystallinity and the orientation requirements. For example, thin ($\sim 1$ μm) (100) GaAs grown on a transparent oxide substrate, such as sapphire, is desired for transmission mode photocathode applications[27, 28]. However, it has not been possible to achieve the (100) GaAs epitaxy on any known sapphire orientation. A further consideration is that a layer as thin as 1 μm grown on a structurally different substrate material may exhibit poor crystallinity. A third buffer layer such as GaP grown between the GaAs film and the sapphire substrate may be feasible since GaP is optically transparent and structurally close to GaAs. The (100) GaP/sapphire epitaxy is achieved[23] through an intermediate thin Si layer ($\sim$2000 Å), and the ultimate multilayer composite substrate is (100) GaP/(100) Si/($1\bar{1}02$) sapphire. (100) GaAs epitaxy has been achieved[21] on this composite substrate.

### *3.2. Substrate crystallinity and surface preparation*

As a general rule in epitaxy, the maximum crystalline perfection which may be achieved in an epitaxial film is of the same order of magnitude as that of the substrate. However, heteroepitaxial materials, especially those grown on foreign substrates with great structural differences, generally exhibit far less crystalline perfection than the substrates.

The dependence of the film crystalline perfection on the substrate bulk crystallinity was studied[11, 23, 29] for GaAs and GaP heteroepitaxial films grown on oxides. Films grown on substrate oxides prepared by different growth methods (Czochralski, flux and Verneuil) exhibit differences in physical appearance. GaAs grown on Verneuil spinel[30] follows the subgrain orientations in the substrates, and films grown on the more perfect Czochralski or flux[31] substrates are more uniformly oriented.

GaAs films grown on spinel substrates prepared by different methods also exhibit a difference in the departure from exact parallel epitaxy with respect to the substrates. This is determined by X-ray diffraction by measuring the angular separation of diffracted peaks from the layer and substrate (as shown in Fig. 4 for GaAs/flux spinel) and comparing the observed value with that expected for parallel epitaxy. GaAs grown on flux spinel substrates of high crystalline perfection (dislocation density $<100$ lines cm$^{-2}$) has the least misorientation, usually less than 0.15°. The GaAs grown on the poorer quality substrates is misoriented by of the order of 0.3° and 0.4° for Czochralski and Verneuil spinel, respectively.

Despite the differences in crystalline properties just described, there is no significant trend of critical dependence of film electrical characteristics on substrate bulk crystalline perfection. This is because the bulk imperfections in films grown on various substrates are of the same order of magnitude (see Section 4). Consequently, factors other than the substrate crystallinity, such as surface quality and impurity contamination, may have a predominant effect on the film electrical characteristics. Results (Table II) obtained from Hall measurements

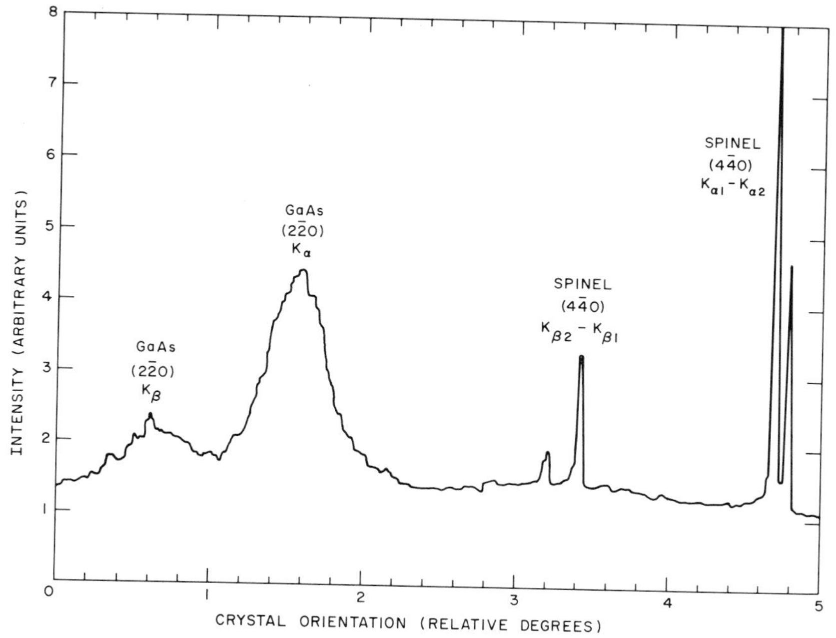


Fig. 4. Trace of diffracted intensity *vs.* crystal orientation for (111) GaAs grown on (111) flux spinel. Peaks from both the substrate and the layer are observed (silver radiation).

TABLE II

ELECTRICAL PROPERTIES OF (111) EPITAXIAL GaAs* GROWN ON SPINEL

| *Substrate* | *Film thickness* (μm) | *Conductivity type* | *Resistivity* (Ω cm) | *Carrier concentration* ($cm^{-3}$) | *Hall mobility* ($cm^2\ V^{-1}\ sec^{-1}$) |
|---|---|---|---|---|---|
| *Flux* $MgAl_2O_3$ | 30 | n | 9 | $9.4 \times 10^{14}$ | 740 |
| *Verneuil* $MgO{:}\,1.7\,Al_2O_3$ | 50 | n | 0.40 | $4.1 \times 10^{15}$ | 3820 |
| *Czochralski* $MgAl_2O_4$ | 23 | n | 0.31 | $4.7 \times 10^{15}$ | 4200 |
| *Czochralski* $MgAl_2O_4$ | 26.4 | p | 0.58 | $3.4 \times 10^{16}$ | 320 |
| *Verneuil* $MgO{:}\,2.0\,Al_2O_3$ | 39 | p | 4.5 | $4.5 \times 10^{15}$ | 316 |

* Unintentionally doped films.

have shown that the electrical characteristics of GaAs grown on Verneuil spinel are comparable with those of GaAs grown on Czochralski spinel. Films grown on flux spinel substrates exhibit generally inferior electrical properties because of substrate impurities[31, 32] that diffuse into the films at the elevated growth temperatures.

Substrate surface preparation is one of the most important growth parameters. The quality of the mechanical polishing has a direct effect on the film growth and electrical characteristics. Surface scratches on the substrates generally provide preferential nucleation sites. Adsorbed layers and impurity aggregates on the substrate surfaces can cause various defects in the epitaxial films. In an attempt to improve the substrate surface quality, the effect of various treatments of the substrate prior to growth has been investigated for the growth of GaAs/oxide[21] and GaP/oxide[11, 23] systems using the organometallic process. The treatments include thermal annealing[4, 22] and chemical polishing[33] of the substrates. However, experimental results to date indicate that films grown on substrates with treatments other than just mechanical polishing followed by standard wet solvent cleaning are generally not superior in quality, and that the growth characteristics are generally not reproducible. These results may not necessarily be fundamental, but represent the present state of the art. Improved substrate surface preparation techniques remain an important objective in the area of heteroepitaxy. The formation of heteroepitaxy using the organometallic growth process in III–V compound/oxide systems without the substrates being annealed and/or etched may possibly be achieved as a result of certain chemical reactions between the constituents in the vapor phase and the substrate surfaces at the growth temperatures, leading to the exposure of single-crystalline substrate surfaces for growth. For example, it was found[21] that $AsH_3$ etches spinel at a temperature of about 700 °C, and the etched single-crystal surfaces promote the epitaxial growth of GaAs on spinel.

### *3.3. Growth temperature, gas flows and source materials*

The optimum growth temperature range for a growth process and film/substrate system depends, to a large extent, on the substrate orientation. The effect of growth temperature on the crystalline quality of heteroepitaxial GaAs on spinel has been studied[21] in the temperature range 500°–800 °C. The general trends observed were: (1) at low temperatures (<600 °C) the GaAs films are polycrystalline in nature; (2) at high temperatures (>800 °C) the gases in the reactor are highly turbulent and the films tend to be inhomogeneous with high densities of growth defects; and (3) within a limited temperature range, 680°–720 °C for (111) spinel and 630°–700 °C for (110) spinel, highly reflective GaAs of good crystallinity can be grown. A similar effect was observed[20] in the epitaxial growth of GaAs on (0001) sapphire in the temperature range 600°–800 °C. Films grown at 800 °C were found to be less perfect than those grown at 675 °C.

The dependence of growth rate on the growth temperature has been studied for both the GaAs/spinel[21] and the GaAs/sapphire[20] systems. In both systems the growth rate was found to be essentially constant over a wide temperature range corresponding to the optimum range for single-crystal growth. The growth rate and the film quality depend largely on the gas flows. The flow conditions that yield the best results for a material system can be determined experimentally from the geometry of the particular apparatus. Generally, higher flow rates than the optimum ones favor heavy deposition at the center of the substrate, while slower rates favor deposition at its periphery.

The growth rate of (111) GaAs on (0001) sapphire by the reaction of $(CH_3)_3Ga$ and $AsH_3$ was found[20] to be essentially linear with $(CH_3)_3Ga$ concentration when the films were deposited in an atmosphere containing As and with at least a tenfold excess of $AsH_3$ over $(CH_3)_3Ga$ in the gas stream entering the reactor. The dependence of growth rate and surface crystallinity on the reactant ratio of $AsH_3$ to $(CH_3)_3Ga$ for the growth of (111) GaAs on (111) spinel is shown in Fig. 5. An optimum range of source gas ratio is shown to be required for the growth of high quality films.

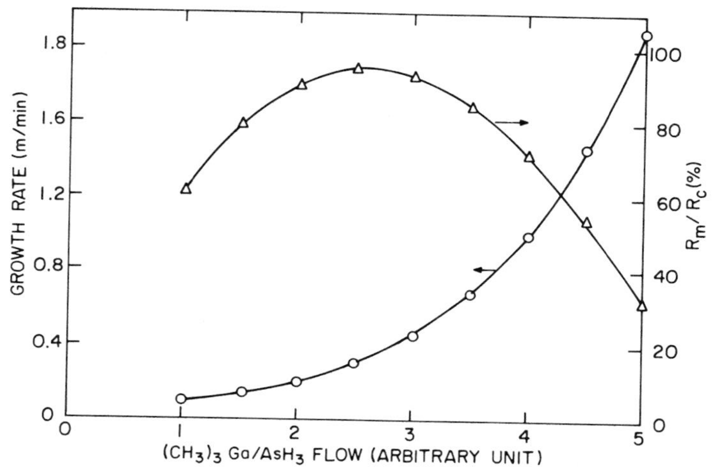


Fig. 5. Growth rate and surface crystallinity as a function of $(CH_3)_3Ga:AsH_3$ flow (surface crystallinity = reflectivity measured/reflectivity calculated).

It has been shown[24] that the electrical characteristics of GaAs-on-sapphire depend largely on the $AsH_3:(CH_3)_3Ga$ flow rate ratios. The carrier concentration of as-grown n-type films has been found to be dependent on the $AsH_3$ flow rate for a given growth temperature and a fixed $(CH_3)_3Ga$ flow rate. For higher $AsH_3$ flow rates the carrier concentration saturates at some value characteristic of the gases used. As the $AsH_3$ flow is reduced, the net donor concentration also tends to decrease. With continued decrease in $AsH_3$ the films eventually become p type. The origin of the acceptor state responsible for the p-type conductivity is probably related to a defect state, possibly an As vacancy. By changing the flow rates of the reactants, p–n junctions have been grown in heteroepitaxial GaAs films[13].

Experimental results obtained from the growth studies have indicated that the purity of the source material is of critical importance in determining the electrical properties of the heteroepitaxial films. Group V hydrides of high purity are commercially available. The purity of the $AsH_3$ gas used for the GaAs growth may be evaluated in a homoepitaxial GaAs growth system employing the reaction between gallium chloride and arsine. However, the quality of the $(CH_3)_3Ga$ varies from lot to lot. The impurities generally found, by emission spectroscopy, in typical lots of $(CH_3)_3Ga$ include Cu, Fe, Zn, Al, Si and Mg of the order $10^{-1}$–$10^2$ ppm (by weight). These impurities may cause significant unintentional doping of the GaAs films. Analysis of $(CH_3)_3Ga$ samples by infrared spectroscopy also revealed the presence of hydrocarbons (degradation products) in widely varying concentrations.

To date $(CH_3)_3Ga$ and other organometallics of controlled quality for the epitaxial growth of compound semiconductors are not commercially available. The nominal high purity materials are generally tested for acceptance by growing the epitaxial films and then measuring the electrical characteristics. GaAs films grown from defective $(CH_3)_3Ga$ exhibit very high resistivity ($>10^3$ $\Omega$ cm) independent of thickness. Sometimes the films are polycrystalline in nature.

## 4. DEFECT CHARACTERIZATION BY X-RAY DIFFRACTION TOPOGRAPHY

### *4.1. Defect structures*

A general and preliminary assessment of crystalline perfection of the heteroepitaxial films grown on oxide substrates may be conveniently made by conventional methods such as Laue X-ray back reflection, glancing angle electron diffraction and various microscope techniques. In addition, the techniques of X-ray diffraction topography are particularly useful in obtaining information on the nature of defect structures in the bulk of films, and sometimes on the effect of substrate defects on the film imperfections. Several III–V epitaxial film/oxide substrate composites have been characterized by topographic methods, with interesting results[11, 23, 29].

The use of Lang transmission topography[34] to characterize heteroepitaxial systems has usually been confined to the examination of layers after removal from the substrate. However, for the III–V compounds (such as GaAs and GaP) grown on sapphire and spinel substrates, the topographic studies can be made on the as-grown sample and allow separate imaging of the layer and substrate. This is due to the high linear X-ray absorption coefficient $\mu$ of the relatively thin GaP or GaAs layer, and the low linear absorption coefficient of the thick substrate. Thus the product $\mu t$ of the linear absorption coefficient and thickness is less than unity for both layer and substrate, and meets the condition $\mu t<1$ for Lang topography using short wavelength (Ag or Mo) X-radiation. Since the two crystalline components of the heteroepitaxial system have different lattice parameters, it is possible to obtain separate topographs of the layer and substrate by appropriate orientation of the composite, first at the Bragg angle for a set of diffracting planes in the substrate, then at the Bragg angle for a set in the layer. Reflection topography can also be used and is, in fact, more convenient (with longer wavelengths, *e.g.* Cu X-radiation) if the layer has a low absorption coefficient or the substrate is highly absorbing.

Examination by X-ray topography reveals that the III–V heteroepitaxial layers, shown to be overall monocrystalline by Laue methods or conventional X-ray diffractometry, are composed of small crystallites ($\sim 10$ µm) most of which are misoriented in a range $\pm 0.1°$ from the nominal orientation of the layer. Further, the general orientation of the layers follows the orientation of large grains in the substrate. Figures 6 and 7 are topographs taken with silver radiation of a (111) GaP film grown on a (0001) sapphire substrate. The two topographs of the sapphire substrate shown in Fig. 6 were taken at crystal settings 0.1° apart and clearly show the presence of at least two subgrains in the crystal. Figure 7 shows topographs obtained from the GaP film, again at a crystal setting difference of 0.1°. It is evident that the GaP layers in Figs. 7(a) and 7(b) follow the orientation of the sapphire grains shown in Figs. 6(a) and 6(b), respectively. Note also in Fig. 6 the clear distinction between the areas where the sapphire grain is present or absent; in Fig. 7 the GaP film orientation is much less sharply defined. For example, in the top part of Fig. 6(a) a sapphire grain exhibits strong diffraction contrast, and there is no diffraction from this same region in Fig. 6(b). The GaP film in the top part of Fig. 7(a) also exhibits strong diffracted contrast, and the image appears to be due to a high density of small crystallites. In Fig. 7(b) this same region is still exhibiting diffracted intensity from a small number of crystallites. Thus the X-ray topographs show that the GaP film is composed of small crystallites which generally follow the orientation of the grains in the substrate, but are misoriented from each other by $\pm 0.1°$.

Topographs of the substrates usually show polishing scratches and defects. The polishing scratches are also visible in topographs of the layers as lines of no

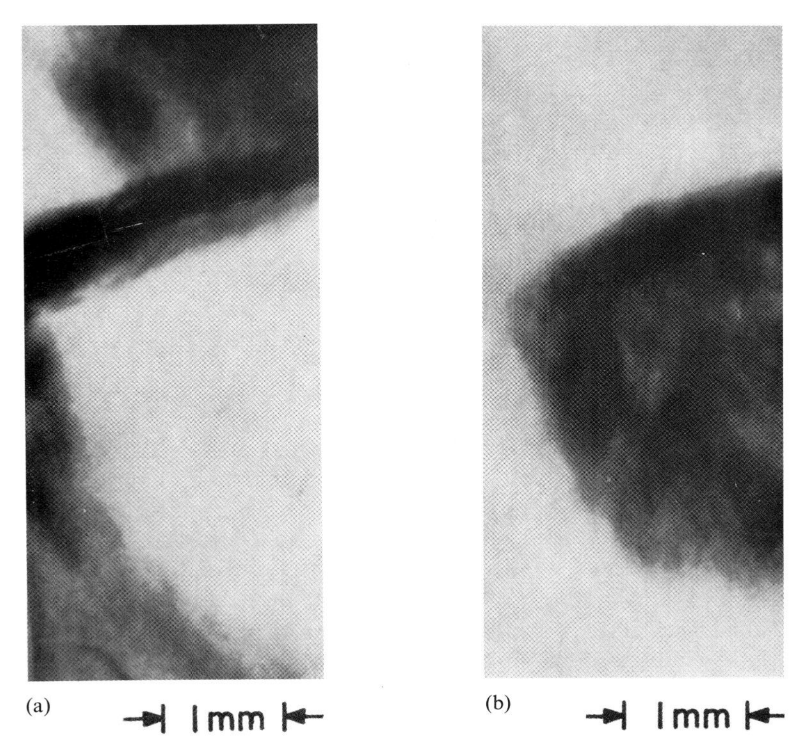


Fig. 6. X-ray transmission topograph of a (0001) sapphire substrate; silver radiation, diffracting from the $(1\bar{2}10)$ planes. The crystal was rotated 0.1° between the two topographs.

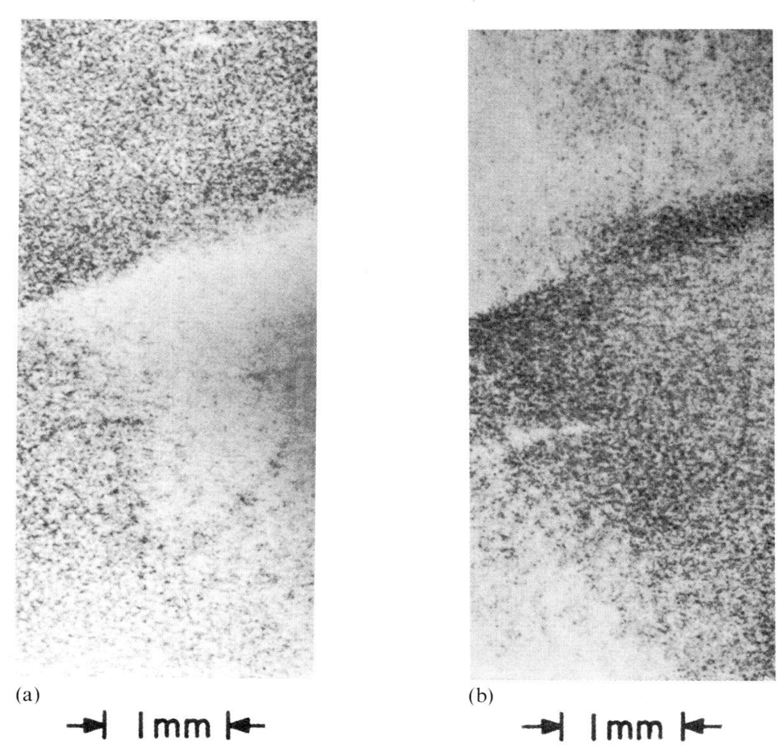


Fig. 7. X-ray transmission topograph of a (111) GaP layer grown on the substrate shown in Fig. 6; silver radiation, diffracting from the (02$\bar{2}$) planes. The crystal was rotated 0.1° between the two topographs.

diffracted intensity, indicating that crystallites grown on the scratches are misoriented from the surrounding layer. Topographs of a (111) GaAs/(111) $MgAl_2O_4$ composite are shown in Fig. 8. The spinel substrate is shown in Fig. 8(a). Several polishing scratches are visible as lines of darker diffracted intensity. Note that there is a uniform background of a dislocation density which is too high ($>10^6$ lines $cm^{-2}$) to resolve individual dislocations. The topograph of the GaAs film is shown in Fig. 8(b). The diffracted image over the entire area of the wafer is evidence that the layer is monocrystalline. However, the dark image is not uniform, and the small whiter areas over the entire topograph are small regions which are misoriented from the nominal crystal orientation. The white lines in the GaAs topographs can be seen to correspond to scratches visible in the topograph of the substrate. This indicates that the GaAs grown on the scratches is misoriented from the rest of the layer. The fact that many of the lines which are visible on the GaAs topographs are not evident on the substrate topograph means that the GaAs growth is sensitive to imperfections (scratches) in the surface of the substrate which do not have a large enough strain field to be imaged by X-ray topography.

The characteristic defect structure in the III–V compounds grown on oxide substrates appears to be inherent in heteroepitaxial composites with great physical,

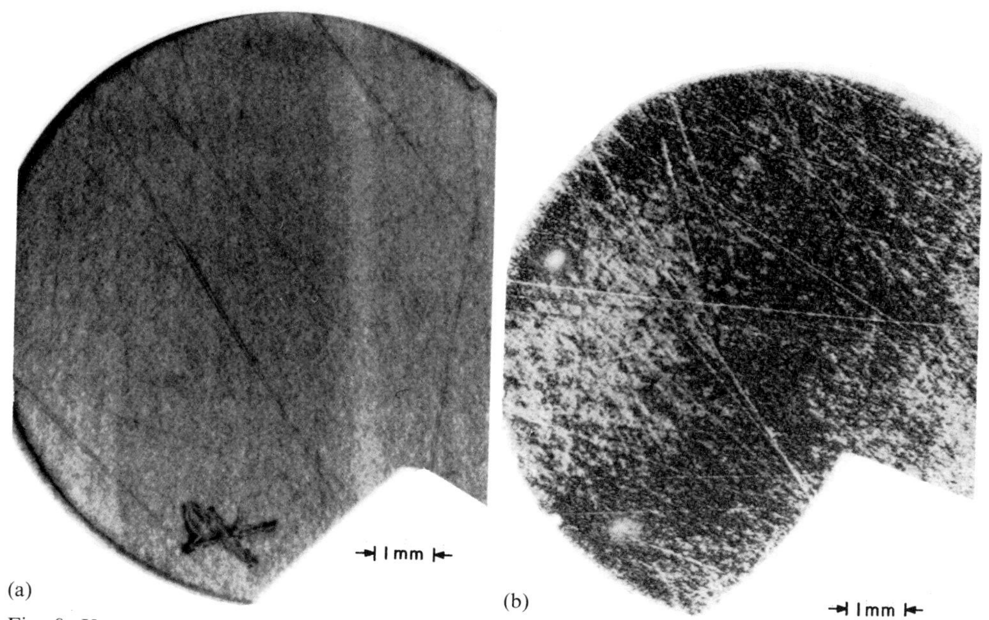


Fig. 8. X-ray transmission topograph of (111) GaAs grown on Czochralski (111) spinel taken with silver radiation: (a) spinel substrate, (440) diffracting planes; (b) GaAs layer 70 μm thick, $(2\bar{2}0)$ diffracting planes.

chemical and structural differences between the layer and the substrate. The defect structure does not appear to depend on the method of epitaxial growth, as similar defects have been observed in films grown by both the OMV (GaAs and GaP) and VPE (GaN) methods. Films grown by LPE on OMV layers also exhibit the characteristic defect structure with some improvement in perfection[11]. Also, LPE (Al, Ga)As films grown on single-crystal GaP are strikingly similar to the III–V/oxide OMV composites[35]. Even epitaxial layers of GaAs grown on Ge by the OMV method (where the layer and substrate have more similar properties) exhibit perfection ranging from small grains (Fig. 9) to better material with a high density of misfit dislocations.

*4.2. Deformation and stress*

The heteroepitaxial films grown on oxide substrates are strained and generally under compressive stress. This is due to the difference in thermal expansion behavior between the film and substrate, and to the elevated growth temperatures. The residual stress may be estimated, to a first order approximation, by the expression[36]

$$\sigma_t = Y_f(\alpha_s - \alpha_f)\,\Delta T \qquad (1)$$

where $\alpha_s$ and $\alpha_f$ are, respectively, the coefficients of thermal expansion of the substrate and the film, $Y_f$ is the elastic modulus of the film, and $\Delta T$ is the temperature change. The thermal expansion coefficient of most III–V semiconductors

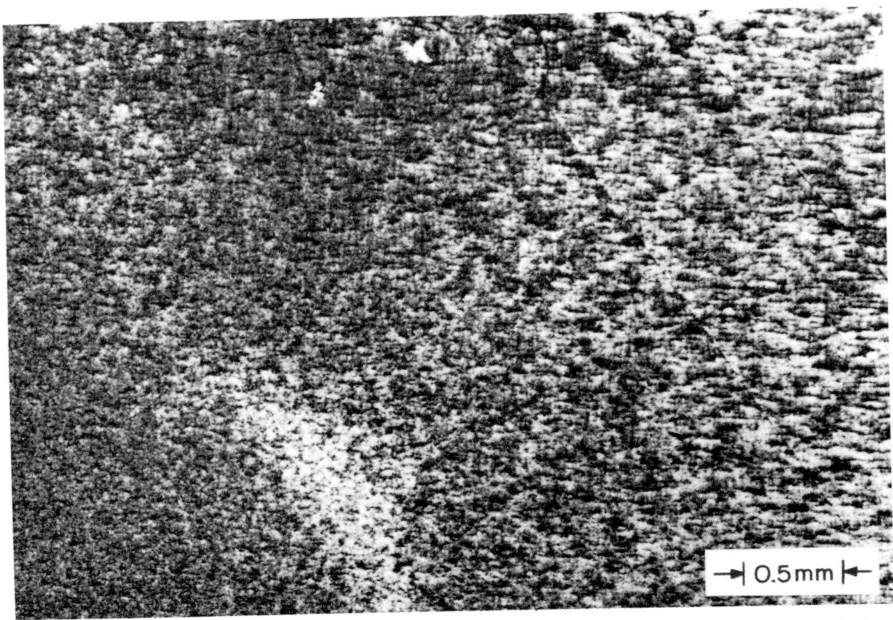


Fig. 9. X-ray reflection topograph of (100) GaAs layer grown on (100) Ge; copper radiation, diffracting from (422) planes.

is less than that of the commonly used oxides such as sapphire, spinel and beryllia. After the film growth, the contraction of the substrate on cooling forces the film into compression and deforms the film/substrate composite, the film being convexed outwards. The deformation and stress may alter some of the physical and electrical properties of a film with respect to the bulk crystalline properties of the same material.

The deformation of a film deposited on a single-crystal substrate can be measured by using an X-ray technique[29] in which the angular setting (measured to seconds of arc) of the substrate crystal corresponding to the Bragg condition for reflection (for maximum $K\alpha_1$ intensity) from selected planes is determined as a function of substrate displacement. Since the structures are curved, the reflecting crystallographic planes are no longer parallel, and the deviation relative to the initial measurement at an edge of the substrate is a function of the horizontal translation of the substrate. The slope of the curve is inversely proportional to the radius of curvature of the crystal. If the curve is a straight line, the bending is uniform across the crystal. The bending is measured about a vertical axis which lies in the reflecting planes and is perpendicular to the direction of translation across the crystal. Thus it is possible to determine if the deformation is isotropic by measuring the bending about various crystallographic directions in the substrate wafer.

The deformation $\delta$ and stress $\sigma$ in an epitaxial film grown on a circular substrate material of radius $r$ may be estimated by the following relations[37]:

$$\delta = r^2/2R \tag{2}$$

and

$$\sigma = E\delta t_s^2/3\,(1-\nu)\,r^2\,t_f \tag{3}$$

where $R$ is the radius of curvature, $t_f$ is the film thickness, $t_s$ the substrate thickness, and $E$ and $\nu$ are, respectively, the elastic modulus and Poisson's ratio of the substrate material.

The deformation of GaAs grown on spinel was determined by the X-ray method. The results are summarized in Fig. 10. The experimentally measured deformation of GaAs on spinel is isotropic as the bending about all measured bending axes is the same. This is expected since crystallographically the GaAs/spinel composite is a cubic-on-cubic system. For a given substrate thickness, the deformation increases with increasing film thickness. The deformation may cause difficulties in device processing and performance. For example, it is known[28] that deformation of GaAs decreases the electron escape probability in negative electron affinity devices. On the other hand, the film crystalline perfection improves as films become thicker. These two factors, the deformation and crystalline perfection, may determine the optimum film thickness at which the best film properties for devices can be achieved. GaAs films up to 70 μm grown on spinel and sapphire substrates ($\sim$20 mil thick) have produced no cracks. The stress in the film, estimated from eqn. (3), is approximately $1 \times 10^9$ dyn cm$^{-1}$ (ref. 29), which is nearly an order of magnitude less than the stress in Si grown on the oxides[22, 37, 38].

The piezoresistance effect (change of resistivity due to isotropic stress) of heteroepitaxial silicon grown on spinel was analyzed theoretically and the results are qualitatively in agreement with experimental observations of the dependence of carrier mobility on the crystallographic orientation of the films. The expected trend of the dependence is that the effective mobility increases and decreases, respectively, for p-(111) and n-(100) silicon films. The piezoresistance effect is

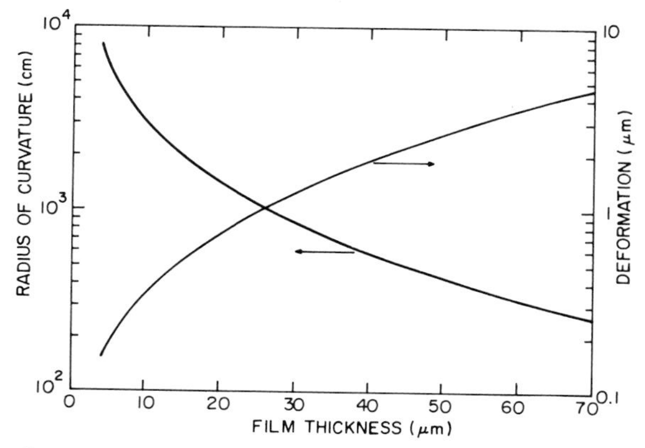


Fig. 10. Radius of curvature and deformation of (111) GaAs on spinel as a function of GaAs thickness (substrate 0.5 mm thick and 0.9 cm in diameter).

less pronounced in the case of GaAs on spinel. This is because the piezoresistance coefficients of GaAs and the film stress are such that the change in resistivity with respect to orientation is minimal. Experimental results indicate that the carrier mobility does not depend significantly on substrate orientation, as in the case of silicon on spinel.

## 5. ELECTRICAL PROPERTIES

Because of the surface defects and bulk imperfections discussed in the preceding sections, it is expected that the electrical properties of heteroepitaxial III–V films grown on oxides are inferior to the bulk material properties. Experimental results obtained to date have been concerned mostly with GaAs films because this material is the most important compound semiconductor and it can be grown relatively easier than many others. Surprisingly, epitaxial GaAs exhibits electrical characteristics favorable enough for potential applications in several devices cited in Section 1 of this paper. The interest in using heteroepitaxial III–V films grown on oxides arises from considerations regarding the substrate size and cost, the dielectric isolation and optical transparency provided by the substrates, the film thermal dissipation, the film–substrate functional coupling, and subsystem integration.

The carrier transport properties of heteroepitaxial GaAs grown (by the organometallic process) on sapphire and on beryllia have been studied in some detail[13, 24]. Properties were reported for films grown on substrate planes that yield the best quality GaAs overgrowths. These include (111) GaAs on (0001) sapphire, (111) GaAs on $(10\bar{1}1)$ beryllia and (100) GaAs on $(11\bar{2}2)$ beryllia. Films grown with no intentional doping on the two substrate materials were found to exhibit, qualitatively, similar behavior in their electrical properties. The first few microns of GaAs grown tend to have high resistivity and exhibit p-type conductivity. With subsequent growth, the outer layers of the film convert to n-type, with donor carrier concentrations ranging from $10^{15}$ to $10^{18}$ $cm^{-3}$ depending largely on the purity of the reactant source materials. The thickness of the underlying high resistivity layer was found to be dependent primarily on the donor carrier concentrations but independent of the total film thickness. Once the film has converted to n-type, the carrier concentration rises and the carrier mobility improves as the film becomes thicker, usually saturating at a film thickness of about 15–20 μm. The exact origins of the acceptor states near the GaAs–oxide interfaces are not known, and several possibilities may exist. These include (1) defects generated as a result of lattice mismatch, thermal contraction differences on cooling from the growth temperatures and/or irregularities in the substrate surface topology, and (2) acceptor impurities either out-diffusing from the substrate or present on the surface at the initiation of growth. The maximum mobilities measured for undoped thick films are about 5000 $cm^2$ $V^{-1}$ $sec^{-1}$ for GaAs/$\alpha$-$Al_2O_3$ and 3000 $cm^2$ $V^{-1}$ $sec^{-1}$ for GaAs/BeO at carrier concentrations in the mid-$10^{16}$ $cm^{-3}$ range. For comparison, homoepitaxial GaAs films grown by the same process were reported to exhibit mobility values of 6900 $cm^2$ $V^{-1}$ $sec^{-1}$ (ref. 39).

Carrier transport properties of GaAs on spinel have been studied[11] and it was found that, in many respects, the properties of undoped (111) GaAs grown on spinel are similar to those of films grown on sapphire and beryllia. Categorically, three kinds of films were obtained under the optimum growth conditions, depending on the impurities present in the growth system. The first class of films exhibits n-type conductivity. A high resistivity ($>10^3\ \Omega$ cm) p layer is often present during the initial growth. With subsequent growth, the material becomes n-type, the carrier concentration rises, and the carrier mobility improves to a certain level. This kind of film exhibits transport properties just like the GaAs/$\alpha$-$Al_2O_3$ and GaAs/BeO films. The highest electron mobility measured is about 5000 $cm^2\ V^{-1}\ sec^{-1}$ at a carrier concentration in the mid-$10^{16}\ cm^{-3}$ range. The second class of films exhibits p-type conductivity. Thick ($>10$ μm) films with mobilities ($\sim 300\ cm^2\ V^{-1}\ sec^{-1}$) close to the bulk value have been prepared in the carrier concentration range mid-$10^{15}\ cm^{-3}$ to mid-$10^{17}\ cm^{-3}$. Since many lots of $(CH_3)_3Ga$ contain appreciable amounts of Cu and Zn which are known to be effective acceptors in GaAs, these p films are most probably Cu- and/or Zn-doped. The third class of films exhibits very high resistivity ($>10^3\ \Omega$ cm) independent of thickness. The films are highly compensated and contain localized p–n junctions.

Undoped (111) GaP epitaxial films grown on sapphire[23] and on spinel[11] exhibit electrical properties with trends similar to those of GaAs grown on oxides. The carrier mobility increases with increasing thickness to about 10 μm, after which it remains essentially constant. The thick films exhibit an overall conductivity type, either n or p, depending mainly on the purity of the source materials. The highest measured mobilities are 80 and 70 $cm^2\ V^{-1}\ sec^{-1}$ for p- and n-type films, respectively. $(11\bar{2}0)$ GaN films grown on sapphire exhibit piezoelectric[26] and electroluminescent[8] properties. Undoped films grown by the organometallic process exhibit mobilities in the range 50–60 $cm^2\ V^{-1}\ sec^{-1}$.

Apart from the cited results, little information is available on the electrical properties of other heteroepitaxial films grown on oxides. This is because many of the films have been prepared only in preliminary experiments. Well-defined materials suitable for electrical measurements are not yet available. It is anticipated that more electrical data will be reported in the literature as more films grown under improved and/or optimum conditions become available.

## 6. CONCLUSIONS

The successful growth of heteroepitaxial III–V compounds on dielectric oxide substrates has been most successfully and reproducibly achieved by the organometallic growth process developed in recent years. The epitaxial film growth characteristics are largely affected by several key parameters including the substrate orientation, substrate surface preparation, growth temperature, gas flow and source material purity. The heteroepitaxial films have been characterized by X-ray diffraction topography and found to be composed of characteristic grain structures which are misoriented by $\pm 0.1°$ from the nominal orientation of the layer. The film/substrate composites are deformed, and the films are generally under compressive stress of the order of $10^9$ dyn $cm^{-2}$. Because of the crystalline defects, it is expected that the electrical properties of the films are inferior to the bulk material properties. Experimental results so far obtained indicate that epitaxial GaAs and GaP films exhibit electrical characteristics favorable enough for device applications in several areas with potential advantages. It is anticipated that more electrical/device data will be reported in the literature as more films grown under improved techniques and/or optimum conditions become available.

## ACKNOWLEDGMENTS

The research reported in this paper was jointly sponsored by the Air Force Materials Laboratory, Air Force Systems Command, Wright-Patterson Air Force Base, Ohio, under Contract No. F33615-70-C-1536, and RCA Laboratories, Princeton, New Jersey.

The authors acknowledge the contributions made by F. C. Dougherty, D. A. Kramer, J. T. McGinn, R. J. Paff, N. Pastal, B. J. Seabury and R. T. Smith for experimental work. They are also grateful to G. W. Cullen, R. E. Honig, H. Kressel, I. Ladany, C. J. Nuese, D. Richman and P. J. Zanzucchi for many helpful and stimulating discussions.

## REFERENCES


1 T. S. La Chapelle, A. Miller and F. L. Morritz, in H. Reiss (ed.), *Progress in Solid State Chemistry*, Vol. 3, Pergamon Press, London, 1967.
2 J. D. Filby and S. Nielson, *Br. J. Appl. Phys.*, *18* (1967) 1357.
3 D. J. Dumin, P. H. Robinson, G. W. Cullen and G. E. Gottlieb, *RCA Rev.*, *31* (1970) 620.
4 G. W. Cullen, *J. Cryst. Growth*, *9* (1971) 107.
5 H. M. Manasevit, *J. Cryst. Growth*, *13–14* (1972) 306.
6 H. M. Manasevit, *J. Cryst. Growth*, *22* (1974) 125.
7 Y. Z. Liu, J. L. Moll and W. E. Spicer, *Appl. Phys. Lett.*, *17* (1970) 60.
8 J. I. Pankove, *J. Lumin.*, *7* (1973) 114.
9 I. Ladany and C. C. Wang, *J. Appl. Phys.*, *43* (1972) 236.
10 I. Ladany and C. C. Wang, *Solid-State Electron.*, *17* (1974) 573.
11 C. C. Wang, I. Ladany, S. H. McFarlane III and F. C. Dougherty, *J. Cryst. Growth*, *24–25* (1974) 239.
12 J. M. Owens, *Proc. IEEE*, *59* (1971) 930.
13 A. C. Thorsen, H. M. Manasevit and R. H. Harada, *Solid-State Electron.*, *17* (1974) 855.
14 J. H. Collins, P. J. Hagon and G. R. Pulliam, *Ultrasonics*, *8* (1970) 105.
15 H. M. Manasevit, *Appl. Phys. Lett.*, *12* (1968) 156.
16 W. M. Yim, E. J. Stofko, P. J. Zanzucchi, J. I. Pankove, M. Ettenberg and S. L. Gilbert, *J. Appl. Phys.*, *44* (1973) 292.
17 H. P. Maruska and J. J. Tietjen, *Appl. Phys. Lett.*, *15* (1969) 327.
18 B. A. Scott, K. H. Nichols, R. M. Potemski and J. M. Woodall, *Appl. Phys. Lett.*, *21* (1972) 121.
19 W. A. Gutierrez, H. D. Pommerrenig and M. A. Jasper, *Solid-State Electron.*, *13* (1970) 1199.
20 H. M. Manasevit and W. I. Simpson, *J. Electrochem. Soc.*, *116* (1969) 1725.
21 C. C. Wang, F. C. Dougherty, P. J. Zanzucchi and S. H. McFarlane III, *J. Electrochem. Soc.*, *121* (1974) 571.
22 C. C. Wang, G. E. Gottlieb, G. W. Cullen, S. H. McFarlane III and K. H. Zaininger, *Trans. Metall. Soc. AIME*, *245* (1969) 441.

23 C. C. Wang and S. H. McFarlane III, *J. Cryst. Growth, 13–14* (1972) 262.
24 A. C. Thorsen and H. M. Manasevit, *J. Appl. Phys., 42* (1971) 2519.
25 H. M. Manasevit, F. M. Erdmann and W. I. Simpson, *J. Electrochem. Soc., 118* (1971) 1864.
26 M. T. Duffy, C. C. Wang, G. D. O'Clock, Jr., S. H. McFarlane III and P. J. Zanzucchi, *J. Electron. Mater., 2* (1973) 359.
27 L. W. James, G. A. Antypar, J. Edgecumbe, R. L. Moon and R. L. Bell, *J. Appl. Phys., 42* (1971) 4976.
28 R. U. Martinelli and D. G. Fisher, *Proc. IEEE, 10* (1974) 1339.
29 S. H. McFarlane III and C. C. Wang, *J. Appl. Phys., 43* (1972) 1724.
30 C. C. Wang, *J. Appl. Phys., 40* (1969) 3433.
31 C. C. Wang and S. H. McFarlane III, *J. Cryst. Growth, 3–4* (1968) 485.
32 C. C. Wang and P. J. Zanzucchi, *J. Electrochem. Soc., 118* (1971) 586.
33 A. Reisman, M. Berkenblit, J. Cuomo and S. A. Chan, *J. Electrochem. Soc., 118* (1971) 1653.
34 A. R. Lang, *J. Appl. Phys., 30* (1959) 1748.
35 M. Ettenberg, S. H. McFarlane III and S. L. Gilbert, in *Proc. 4th Int. Symp. on GaAs and Related Compounds*, Conf. Ser. No. 17, Inst. Phys., London, 1972.
36 Y. Budo and J. Priest, *Solid-State Electron., 6* (1963) 159.
37 D. J. Dumin, *J. Appl. Phys., 36* (1965) 2700.
38 C. Y. Ang and H. M. Manasevit, *Solid-State Electron., 8* (1965) 994.
39 S. Ito, T. Shinohara and Y. Seki, *J. Electrochem. Soc., 120* (1973) 1419.

*Thin Solid Films*, *31* (1976) 25–37
© Elsevier Sequoia S.A., Lausanne—Printed in Switzerland


# KINETICS OF THE HYDROGEN EFFECT ON PbSe EPITAXIAL FILMS

J. J. YOUNG* AND J. N. ZEMEL
*Department of Electrical Engineering and Science, Moore School of Electrical Engineering, University of Pennsylvania, Philadelphia, Pa. 19174 (U.S.A.)*
(Received May 23, 1975; accepted June 17, 1975)

A one-dimensional diffusion equation has been solved to establish the kinetic behavior of atomic hydrogen in PbSe under both steady state a.c. hydrogen excitation and initial step conditions. The boundary conditions on the surfaces of the films were constructed with the surface recombination velocity $g$ as a phenomenological parameter. Since the hydrogen effect makes the film more n-type, measurements of electrical conductivity changes were used to determine the amount of atomic hydrogen inside the film. Step and periodic hydrogen flux–time variations were used to determine $g$ and the diffusion coefficient $D$, respectively. $g$ was linearly proportional to the partial oxygen pressure. The activation energy for H diffusion was 0.4 eV and the pre-factor was $1.1 \times 10^{-7}$ cm$^2$ sec$^{-1}$ for the p-type films measured. The assumed boundary conditions for the analysis were established as a result of the experimental measurements. The amount of atomic hydrogen that diffused into the conventional vacuum evaporated p-type epitaxial PbSe films was much larger than that which diffused into an n-type epitaxial film grown by the hot wall method. This result strongly suggests that the hot-wall-grown film had fewer Pb vacancies than the vacuum-evaporated films.

## INTRODUCTION

The effect of ambient gases on the electrical properties of epitaxial lead chalcogenide films has been investigated by many authors. It was observed by Brodsky and Zemel[1], in 1967, that oxygen would adsorb on lead selenide films driving the surface p-type, and that nitrogen, helium and argon had no observable effect on the electrical properties. They also reported that a gas generated by the hot filament of an ionization gauge reversed the effect of the oxygen.

Egerton and Juhasz[2] showed that there were long-term and short-term oxygen effects on PbTe films. The former was due to oxygen diffusion and was irreversible while the latter was caused by oxygen surface adsorption. The gas responsible for the reversal of the oxygen effect on PbSe was identified by McLane

---

* Present address: RCA Bloomington, Indiana, U.S.A.

and Zemel[3] as atomic hydrogen dissociated from the hydrocarbons in the background diffusion pump vapor by the filament. Egerton and Crocker[4] raised the possibility of unintentional doping of the PbTe films with atomic hydrogen generated by the high temperature sources. This possibility would account for part of the long-term irreversible oxygen effect. A more systematic study of ambient gas effects on PbSe epitaxial films was conducted by McLane[5]. His results showed that the excess carrier concentration varied as the square-root of time during exposure to atomic hydrogen, strongly indicating a diffusion phenomenon. He also studied the time dependence of the oxygen effect and found it to be logarithmic. This logarithmic time dependence had been discussed by Taylor and Thon[6] for surface chemisorption. In papers both by McLane and Zemel[3] and by Egerton and Crocker[4], the mechanism of atomic hydrogen reacting with surface-adsorbed oxygen to form water was suggested.

Based upon these facts, a mathematical model can be constructed assuming that atomic hydrogen can diffuse into the film bulk while oxygen atoms adsorb on the surface of the PbSe films. We have investigated an experimental method for measuring the atomic hydrogen diffusion constant as well as its temperature dependence, and have determined the role of surface chemisorbed oxygen in reversing the effect of atomic hydrogen.

In this work, p-type PbSe epitaxial films were used because they are much more sensitive to atomic hydrogen than n-type films. Since the conductance of PbSe films is sensitive to the presence of atomic hydrogen, these films show promise as atomic hydrogen detectors.

## MATHEMATICAL MODEL

In a PbSe film the atomic hydrogen occupies a lead vacancy site and eliminates one of the holes associated with the site. By defining the number of hydrogen atoms at lead vacancy sites and the concentration of hydrogen atoms as $n_H$ and $N_H$ respectively, McLane[5] has proposed that

$$n_H = \beta N_H \tag{1}$$

The constant $\beta$ defined in this way is the probability of finding a hydrogen atom at a lead vacancy site.

Since the hydrogen effect is a diffusion phenomenon, both $N_H$ and $n_H$ obey Fick's second law:

$$D\nabla^2 n_H = \partial n_H / \partial t \tag{2}$$

where $D$ and $t$ are the diffusion constant and time respectively. For a homogeneous film of thickness $d$, width $w$ and length $l$, if the effect of the surface on carrier scattering is small and can be neglected, the conductance $G_0$ can be expressed as

$$G_0 = \frac{w}{l}\int_0^d \sigma \mathrm{d}z = \frac{w}{l} e\mu_p \int_0^d p(z)\,\mathrm{d}z \tag{3}$$

where $e$ is the electronic charge, $\mu_p$ is the hole mobility and $\sigma$ is the conductivity. When atomic hydrogen diffusion is taking place, the change in the concentration of electrons per unit volume is $n_H(z, t)$. Let $\Delta G(t)$ be the change in conductance of the film, and let the condition $p(z) - n_H(z, t) > 0$ be satisfied for all values of $z$ and $t$. Then the following expressions can be obtained:

$$G(t) = G_0 - \Delta G(t)$$
$$\Delta G(t) = -\frac{w}{l} e\mu_p \int_0^d n_H(z, r)\, dz \tag{4}$$

Since the atomic hydrogen diffusion is in the direction perpendicular to its surface, a one-dimensional diffusion equation can be solved to analyze this phenomenon:

$$D\, \partial^2 n_H/\partial z^2 = \partial n_H/\partial t \tag{5}$$

with the initial condition

$$n_H = 0 \quad \text{for} \quad t \leqq 0 \tag{6}$$

$z = 0$ is defined as the interface between the substrate and the PbSe epitaxial film. For reasons which will be discussed later, the flux of atomic hydrogen at this interface is zero. From Fick's first law, the one-dimensional flux can be expressed with the boundary condition

$$J_H(z, t) = -D\, \partial n_H/\partial z$$
$$J_H(0, t) = 0 \tag{7}$$

Let $\Gamma_H(t)$ be the atomic hydrogen flux in space and $s$ the effective sticking coefficient. The incoming flux at the front surface ($z = d$) is then $s\Gamma_H$. Two assumptions have to be made for the outgoing flux. These two assumptions will be discussed later and can be proved to be valid by the experimental results of this paper. First, the flux is assumed to be proportional to the atomic hydrogen concentration in the surface region $N_H(d, t)$, or equivalently $n_H(d, t)$, with a proportionality constant $g$ defined as the surface recombination velocity. $g$ has the units of velocity, cm sec$^{-1}$. Second, $g$ is assumed to be independent of $N_H(d, t)$, or equivalently $n_H(d, t)$. Therefore the net flux at $z = d$ is the difference of these two terms:

$$J_H(d, t) = s\Gamma_H(t) - g n_H(d, t) \tag{8}$$

The surface atomic hydrogen recombination velocity is an important phenomenological parameter and is investigated in this work. It is experimentally possible to set $\Gamma_H(t)$ equal to zero. Since $g$ is a constant, eqn. (8) requires that $n_H(d, t)$ also becomes a constant whose value is determined by the value of $\Gamma_H(t)$:

$$J_H(d, t) = 0$$
$$n_H(d, t) = (s/g)\Gamma_H(t) = \text{constant} \tag{9}$$

After $n_H(d, t)$ attains a constant value $n_H(d, t_0)$, $\Gamma_H(t)$ is suddenly set equal to zero at time $t = t_0$. Then eqn. (8) becomes

$$J_H(d, t_0) = gn_H(d, t_0) \tag{10}$$

It can be shown from eqns. (4) and (10) that

$$\frac{\mathrm{d}}{\mathrm{d}t}\{\Delta G(t)\}_{t=t_0} = -\frac{we\mu_p}{l}\, gn_H(d, t_0) \tag{11}$$

$$G(t_0) = G_0 - (we\mu_p/l)\, n_H(d, t_0)\, d \tag{12}$$

For another value of $\Gamma_H$ this procedure leads to another set of equations for $\Gamma_{Hn}$ and $t = t_n$, $n = 1,2,3,\ldots$:

$$\begin{aligned} \frac{\mathrm{d}}{\mathrm{d}t}\{\Delta G(t)\}_{t=t_n} &= -\frac{we\mu_p}{l}\, gn_H(d, t_n) \\ G(t_n) &= G_0 - (we\mu_p d/l)\, n_H(d, t_n) \end{aligned} \tag{13}$$

Since the absolute value of $n_H(d, t_n)$ cannot be obtained, at least two measurements are needed in order to determine $g$. Let $m$ be any positive integer which is not equal to $n$. Using eqns. (13) we obtain

$$\frac{\mathrm{d}}{\mathrm{d}t}\{\Delta G(t)\}_{t=t_n} - \frac{\mathrm{d}}{\mathrm{d}t}\{\Delta G(t)\}_{t=t_m} = \frac{g}{d}\{G(t_n) - G(t_m)\} \tag{14}$$

This equation was used to determine $g$ in our experiments.

A square-wave excitation of period $T_0$ can be generated experimentally for the function $\Gamma_H(t)$ as shown in Fig. 1. If this $\Gamma_H(t)$ has been applied to a film for a very long period, a steady state can be reached. The steady state solution of $n_H(z, t)$ can be solved by Fourier analysis. This type of derivation can be found in a paper by Van der Ziel[8]. A complete derivation is given in the Appendix. Only the important equations for one special case will be discussed here.

The solution of the high frequency approximation at the fundamental frequency $\omega/2\pi$ is

$$\Delta G_1 = -\frac{we\mu_p s\Gamma}{l}\,\frac{4L_H}{\pi}\left\{1 + j\left(1 + \frac{\sqrt{2\omega D}}{g}\right)\right\}^{-1} \tag{15}$$

where $\Delta G_1$ is the change of conductance at the fundamental frequency and $\theta$ is the phase angle between the function $\Gamma_H$ and $\Delta G_1$. The high frequency restriction is

$$\omega \gg D/2d^2 \tag{16}$$

The solution of the low frequency approximation is not given here because low frequency requires an experimentally impractical frequency—much smaller than $10^{-5}$ Hz for values of $D$ at room temperature and a film thickness of 1000 Å.

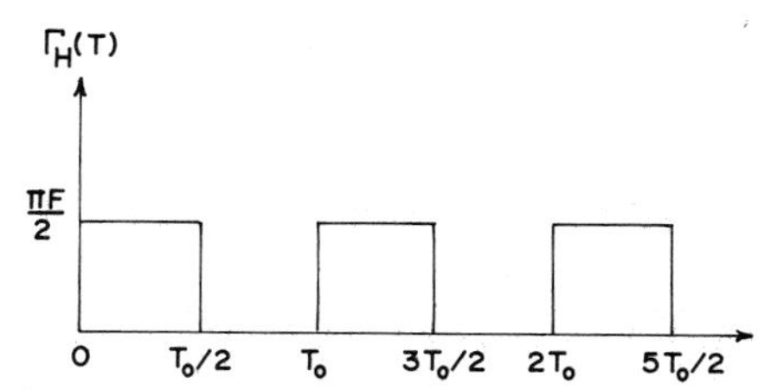


Fig. 1. Square-wave excitation function of $\Gamma_H(t)$.

## EXPERIMENTAL RESULTS

The value of $g$ for atomic hydrogen was measured in the presence of oxygen. This was done by recording the resistivity voltage of the sample on a chart recorder. After the vacuum system had been completely evacuated to a residual pressure $p_r$, the hydrogen leak valve was set with an ion gauge so that the additional molecular hydrogen pressure $p_{H_2}$ was $10^{-5}$ Torr. Then the oxygen leak valve was set to a desired pressure $p_{O_2}$. The total pressure was $p_r + p_{H_2} + p_{O_2}$. The next step was to set the filament current to an appropriate value. Usually the Hall voltage was measured before the $g$ measurement. Then, by suddenly turning off the filament current at time $t_0$, the sample voltage began to decrease as shown in Fig. 2. By graphically measuring the initial slope, $g$ can be calculated from the measurements carried out at different $\Gamma_H$ values. In this fashion, by varying $p_{O_2}$, the relation between $g$ and $p_{O_2}$ can be obtained using eqn. (14). Measurements at different temperatures yielded an activation energy $E_g$ of $g$.

The next step was to measure the diffusion coefficient for hydrogen diffusion. After setting $p_{O_2}$ and $p_{H_2}$ to desired levels, the filament current was set to achieve a desired steady state $IR$ drop $V_R$. By turning the filament on and off with a switch to generate a square-wave flux of atomic hydrogen, a steady state was reached. The response of $V_R$ to this excitation was obtained on the chart recorder as shown in Fig. 3. From the $V_R$ waveform, the phase angle of the fundamental frequency with respect to the excitation can be extracted by Fourier analysis. With a known value of $g$ and this phase angle, the hydrogen diffusion constant can be calculated using eqn. (15). The diffusion activation energy $E_D$ can be calculated from the temperature dependence of $D$. As an example, data from sample A are presented in Fig. 4 for different oxygen partial pressures $p_{O_2}$, where $G_n$ is defined as $\Delta G(t_n)$. The conditions under which the data were taken are specified in this figure. Data from five other samples all strongly support the linear relation between $\mathrm{d}G_n/\mathrm{d}t$ and $G_n$ for all samples measured. From the experimental results and eqn. (14), we conclude that $g$ is indeed independent of $\Gamma_H$, and that the boundary condition of eqn. (8) is valid. The values of $g$ are plotted against different oxygen partial pressures in Fig. 5 for sample A at several temperatures. The fact that the lines do not pass through the origin is due to the

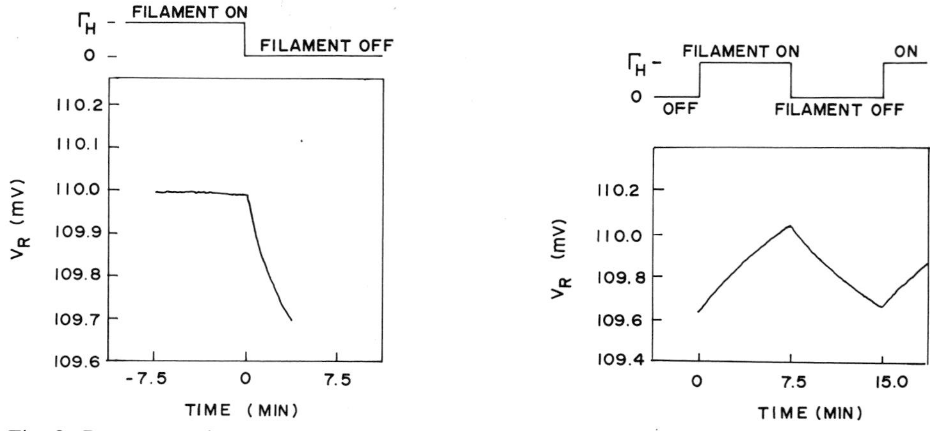


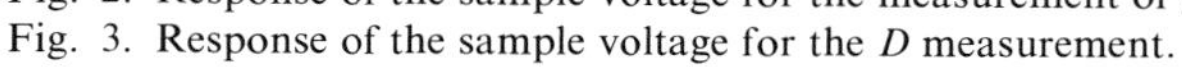
Fig. 2. Response of the sample voltage for the measurement of $g$.
Fig. 3. Response of the sample voltage for the $D$ measurement.

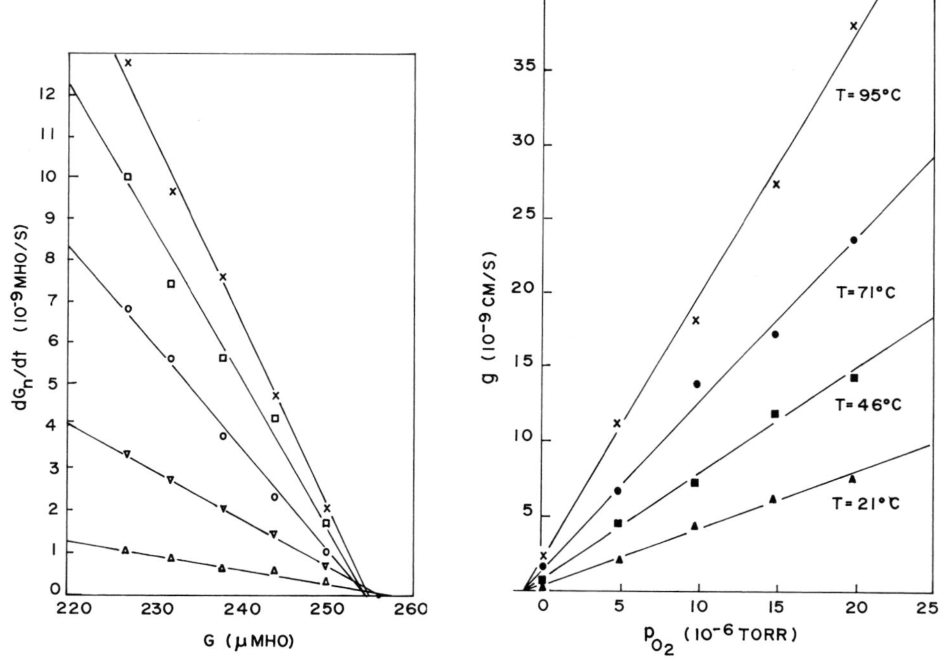


Fig. 4. $dG_n/dt$ *vs.* $G_n$ for sample A at 21 °C with $p_r = 5\times10^{-6}$ Torr and $p_{H_2} = 10\times10^{-6}$ Torr.
Fig. 5. $g$ *vs.* $p_{O_2}$ for sample A with $p_r = 5\times10^{-6}$ Torr and $p_{H_2} = 10\times10^{-6}$ Torr.

residual gas pressure in the vacuum. Temperature instability was a problem during these measurements and led to some scatter. It can be estimated from this figure that the oxygen partial pressure of the residual gases is approximately $1.5\times10^{-6}$ Torr, in excellent agreement with the residual pressure. For $D$ measurements, the fundamental angular velocity $\omega$ is equal to $2\pi/T_0$. At a particular temperature with a preset oxygen partial pressure $p_{O_2}$, $D$ and $g$ are constants. Therefore a plot of tan $\theta$ against $\omega$ for several values of $\omega$ should be a straight line passing through the point $\tan\theta = 1$ at $\omega = 0$ provided that the high frequency approximation is valid.

Experimental values of tan $\theta$ were obtained from a Fourier analysis of the data. Examples are shown in Fig. 6 for a number of samples as functions of $\sqrt{\omega}$. These experimental curves show excellent agreement with eqn. (5) and are strong evidence for the validity of the boundary conditions.

It is obvious that the diffusion constant can be calculated from eqn. (15) when tan $\theta$ and $g$ are measured and $\omega$ is known. The measured diffusion constants are listed in Table I for all six samples.

TABLE I

HYDROGEN DIFFUSION CONSTANTS ($cm^2\ sec^{-1}$)

| *Sample* | *T* = *21 °C* | *46 °C* | *71 °C* | *96 °C* |
|---|---|---|---|---|
| A | $1\times10^{-14}$ | $4.8\times10^{-14}$ | $10\times10^{-14}$ | $25.5\times10^{-14}$ |
| B | $1.4\times10^{-14}$ | – | $14\times10^{-14}$ | – |
| C | $4.2\times10^{-14}$ | – | $45\times10^{-14}$ | – |
| D | $1.2\times10^{-14}$ | – | $9.6\times10^{-14}$ | – |
| E | $1.2\times10^{-14}$ | – | $10.4\times10^{-14}$ | – |
| F | $2.2\times10^{-14}$ | – | – | – |

Although there is no other source of information concerning the diffusion constant of hydrogen in PbSe films, a rough estimate by Egerton and Crocker[4] for the diffusion constant of atomic hydrogen in PbTe films at 170 °C gave the value of $10^{-11}$ $cm^2\ sec^{-1}$. By assuming that the diffusion constant of hydrogen in PbSe is approximately the same and using the diffusion constant activation energy of 0.4 eV obtained by McLane[5], we calculated that $D$ at 21 °C is approximately $5\times10^{-14}$ $cm^2\ sec^{-1}$. This value is quite close to the values listed in Table I. It is well known that the diffusion constant varies exponentially with $1/T$ and has an activation energy[11]

$$D = D_0 \exp(-E_D/kT) \tag{18}$$

where $E_D$ is the activation energy for diffusion and $D_0$ is an experimentally determined constant with a value of $1.1\times10^{-7}$ $cm^2\ sec^{-1}$. A plot of log $D$ against $1/kT$ is shown in Fig. 7. $E_D$ can easily be obtained from these figures and values are listed in Table II for the five samples measured. These values compare favorably with the values obtained by McLane[5].

The temperature range used in this study was from 21 °C to 96 °C. At higher temperatures, measurement uncertainty caused by oxygen diffusion and/or PbSe sublimation results in large measurement errors. At lower temperatures, the time to reach a steady state becomes quite long due to the decrease in the diffusion coefficient. As a result, extremely good temperature stability must be maintained for a period of hours. The surface recombination phenomenon is

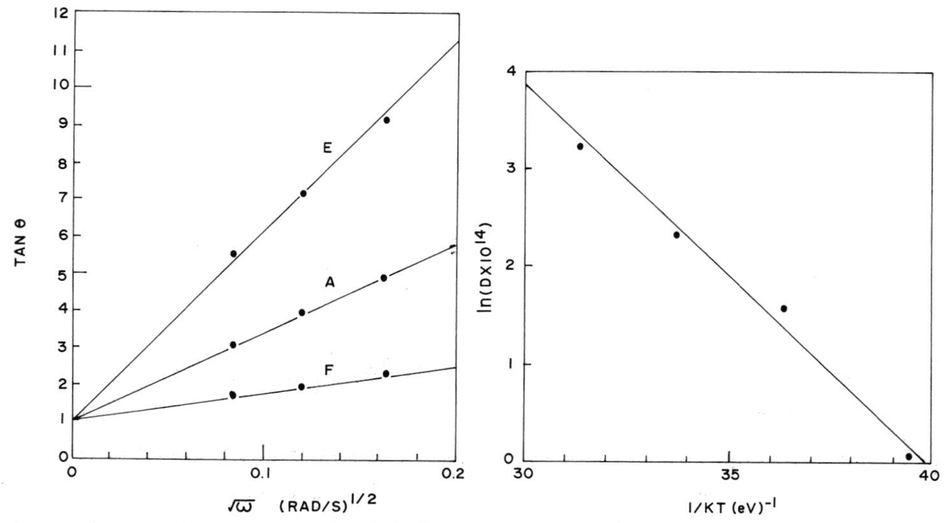


Fig. 6. Tan $\theta$ *vs.* $\sqrt{\omega}$ for three samples at 21 °C with $p_r = 5 \times 10^{-6}$ Torr and different oxygen pressures $p_{O_2}$:E, $10 \times 10^{-6}$ Torr; A, $15 \times 10^{-6}$ Torr; F, $20 \times 10^{-6}$ Torr.
Fig. 7. Ln $(D \times 10^{14})$ *vs.* $1/kT$ for sample A.

TABLE II

ACTIVATION ENERGY OF $D$

| *Sample* | A | B | C | D | E |
|---|---|---|---|---|---|
| $E_D$(eV) | 0.4 | 0.4 | 0.41 | 0.36 | 0.38 |

due to the reaction between atomic hydrogen and oxygen. In addition, oxygen surface adsorption is also involved. The temperature dependences of these processes have an Arrhenius form $\exp(-E_g/kT)$, where $E_g$ is the activation energy[9, 10]. Thus

$$g = g_0 \exp(-E_g/kT)$$

where $g_0$ is a constant and can be determined experimentally. Replotting Fig. 5 indicates that the data follow the Arrhenius form, as shown in Fig.8. It is apparent that the activation energy is independent of oxygen partial pressure within experimental accuracy. Values for the five samples measured are listed in Table III.

TABLE III

ACTIVATION ENERGY OF $g$

| *Sample* | A | B | C | D | E |
|---|---|---|---|---|---|
| $E_g$(eV) | 0.2 | 0.21 | 0.2 | 0.24 | 0.23 |

Although the activation energy of $g$ can be measured, it is still far from clear which process is rate limiting. The difficulty is that many processes are involved in the recombination of H with oxygen on the PbSe surface and these cannot be separately identified.

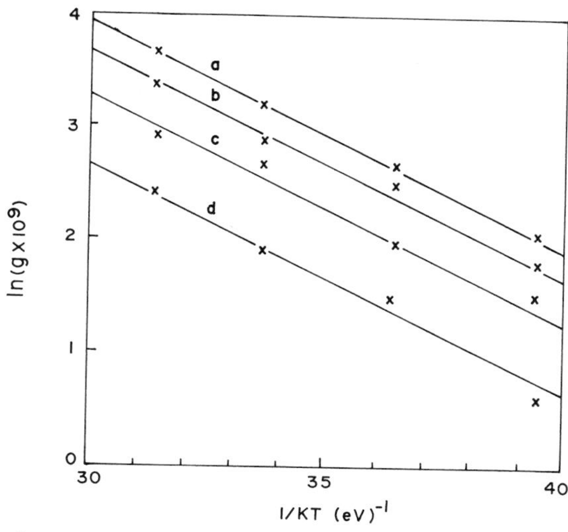


Fig. 8. Ln ($g \times 10^9$) *vs.* $1/kT$ for sample A.

## DISCUSSION

In order to obtain accurate measurements, the conductance change due to the hydrogen effect is limited to the linear region of the corresponding Hall coefficient *versus* resistivity plot. This ensures that there is only one type of carrier throughout the film. Therefore the change of conductance is always proportional to the total number of majority carriers. This can be considered as an extension of the Petritz model[12].

The most important factor affecting measurements is temperature stability. This can be understood more clearly by citing some important numerical quantities. If $V_R$ is 200 mV at 21 °C, it will rise to approximately 350 mV at 96 °C due to the effect of temperature on mobility. This is equivalent to 2 mV °C$^{-1}$. In our measurements the excitation function had a period of 15 min and could be titled 0.2 mV over the time of excitation if there was a temperature drift of 0.1 °C in 15 min. The actual tilt in our experiment corresponds roughly to 0.06 °C drift in 15 min. The displacement of the conductivity was only 0.4 mV. Therefore drift was a problem of some consequence. It is quite difficult to maintain a temperature stability of $10^{-2}$ °C h$^{-1}$. For measurements at higher temperatures, it becomes even more difficult to maintain temperature stability because of the long heat conduction paths and heater resistance variation. The next important factor is the stability of the leak valves and the reading accuracy of the ion gauge. The other sources of inaccuracy in the measurements are much smaller relative to the two cited above. A tilt correction has been included in the Fourier series waveform analysis for all calculations for $D$.

Because $g$ is proportional to the oxygen partial pressure, it can be concluded that atomic hydrogen recombines with oxygen on the PbSe film surface. The method developed to measure the diffusion constant of atomic hydrogen yields accurate values of this phenomenological parameter. These measurements provide further evidence on the diffusion of atomic hydrogen in lead selenide. The experimental results strongly support the validity of the two boundary conditions for the diffusion equation, *i.e.* the impenetrable barrier at $z = 0$ and the radiation boundary condition at $z = d$.

The data in Fig. 5 indicate that $g = 0$ when $p_{O_2} = 0$. If any H entering the film diffused through the PbSe–NaCl interface, there would be an effective $g$ associated with that interface. The conductivity would then drift at very low oxygen pressures at a fixed rate. This was not observed. Therefore the impenetrable barrier at $z = 0$ appears to be valid and eqn. (7) is justified. Since $g$ is independent of $n_H(d, t)$, the boundary condition of eqn. (8) is also valid. A summary of the experimental results is presented in Table IV.

The overall evidence strongly suggests that the conductivity change is due to hydrogen atoms at lead vacancy sites[5, 13] only. Thus only films with appreciable lead vacancies can be measured by our experiment. Under the same condition of atomic hydrogen generation, *i.e.* the same $H_2$ partial pressure and the same current in the tungsten filament, the initial slope of the conductance change was used to calculate the amount of atomic hydrogen diffusing into the films. For a conventional vacuum-evaporated p-type epitaxial PbSe film, the rate was $4 \times 10^{-10}$ atoms $cm^{-2}$ $sec^{-1}$. It would be reasonable to expect that p-type films have more lead vacancies. For the one hot wall epitaxial n-type PbSe film examined, the slope was too low to be measured accurately. This suggests that the number of Pb vacancies in this film is substantially lower than in the conventional n-type films measured by McLane and Zemel[3]. This result is in keeping with other measurements on hot-wall-grown epitaxial PbSe films[16].

We conclude that the hydrogen effect in lead salts provides an extraordinary means of monitoring Pb vacancies in a sample. However, in order to determine quantitatively the relation between atomic hydrogen diffusion and lead vacancies, further work must be done.

ACKNOWLEDGMENTS

This work was performed in partial satisfaction of the requirements for a Ph.D. in Electrical Engineering.

The authors would like to express their appreciation to A. Farhat for the growth of the films used in our experiment and to the Leeds and Northrup Company for generous assistance and support to one of us (J.J.Y.).

REFERENCES

1 M. H. Brodsky and J. N. Zemel, *Phys. Rev.*, *155* (1967) 780.
2 R. F. Egerton and C. Juhasz, *Thin Solid Films*, *4* (1969) 239.
3 G. McLane and J. N. Zemel, *Thin Solid Films*, *7* (1971) 229.
4 R. F. Egerton and A. J. Crocker, *Surf. Sci.*, *27* (1971) 117.

TABLE IV

TABULATED RESULTS

| | | | Sample A | B | C | D | E | F |
|---|---|---|---|---|---|---|---|---|
| $\mu$ at 21 °C ($cm^2\ sec^{-1}\ V^{-1}$) | | | 610 | 840 | 550 | 545 | 470 | 910 |
| $p$ ($cm^{-3}$) | | | $1.2 \times 10^{18}$ | $2 \times 10^{18}$ | $2 \times 10^{18}$ | $1.5 \times 10^{18}$ | $2.5 \times 10^{18}$ | $1.5 \times 10^{18}$ |
| $g$ ($cm\ sec^{-1}$) | 21 °C | $p_{O_2} = 1 \times 10^{-5}$ Torr | $4.4 \times 10^{-9}$ | $5.3 \times 10^{-9}$ | $5.0 \times 10^{-9}$ | $2.0 \times 10^{-9}$ | $2.5 \times 10^{-9}$ | – |
| | | $p_{O_2} = 2 \times 10^{-5}$ Torr | $7.5 \times 10^{-9}$ | $8.9 \times 10^{-9}$ | $9.5 \times 10^{-9}$ | $3.8 \times 10^{-9}$ | $4.8 \times 10^{-9}$ | $27.5 \times 10^{-9}$ |
| | 71 °C | $p_{O_2} = 1 \times 10^{-5}$ Torr | $13.8 \times 10^{-9}$ | $17 \times 10^{-9}$ | $16.1 \times 10^{-9}$ | $8.5 \times 10^{-9}$ | $10.1 \times 10^{-9}$ | – |
| | | $p_{O_2} = 2 \times 10^{-5}$ Torr | $23.6 \times 10^{-9}$ | $32 \times 10^{-9}$ | $29.4 \times 10^{-9}$ | $15.2 \times 10^{-9}$ | $16.5 \times 10^{-9}$ | – |
| $E_g$ (eV) | | | 0.2 | 0.21 | 0.2 | 0.24 | 0.23 | – |
| $D$ ($cm^2\ sec^{-1}$) | 21 °C | | $1 \times 10^{-14}$ | $1.4 \times 10^{-14}$ | $4.2 \times 10^{-14}$ | $1.2 \times 10^{-14}$ | $1.2 \times 10^{-14}$ | $2.2 \times 10^{-14}$ |
| | 71 °C | | $10 \times 10^{-14}$ | $14 \times 10^{-14}$ | $45 \times 10^{-14}$ | $9.6 \times 10^{-14}$ | $10.4 \times 10^{-14}$ | – |
| $E_D$ (eV) | | | 0.4 | 0.4 | 0.41 | 0.36 | 0.38 | – |

5 G. McLane, *Ph.D. Dissertation*, Univ. of Pennsylvania, Phila., Pa., 1971.
6 H. A. Taylor and N. Thon, *J. Am. Chem. Soc.*, *74* (1952) 4169.
7 A. S. Grove, *Physics and Technology of Semiconductor Devices*, Wiley, London, 1967.
8 A. Van der Ziel, *J. Appl. Phys.*, *44* (2) (1973) 546.
9 K. Denbigh, *The Principles of Chemical Equilibrium*, Cambridge Univ. Press, Cambridge, 1968.
10 J. R. Anderson, *Chemisorption and Reactions on Metallic Films*, Academic Press, London, 1971.
11 A. J. Dekker, *Solid State Physics*, Prentice-Hall, Englewood Cliffs, N.J., 1957.
12 R. L. Petritz, *Phys. Rev.*, *110* (1958) 1254.
13 N. J. Parada and G. W. Pratt, Jr., *Phys. Rev. Lett.*, *22* (5) (1969) 180.
14 H. Hochstadt, *Differential Equations*, Holt, Rinehart and Winston, New York, 1964.
15 H. S. Carslaw and J. C. Jaeger, *Conduction of Heat in Solids*, Oxford Univ. Press, London, 1959.
16 P. Carmichael, Personal communication, 1974.

## APPENDIX

*Fourier analysis of the one-dimensional diffusion equation*

Since eqn. (7) is a linear differential equation, the separation of variables method[14] can be used to solve for a steady state solution of the one-dimensional diffusion equation:

$$D\,\partial^2 n_H/\partial z^2 = \partial n_H/\partial t \tag{A1}$$

$$J_H(z, t) = -D\,\partial n_H/\partial z \tag{A2}$$

with boundary conditions

$$\begin{aligned} &J_H(0, t) = 0 \\ &J_H(d, t) = s\Gamma_H(t) - g n_H(d, t) \end{aligned} \tag{A3}$$

The Fourier series expansion for the waveform in Fig. 1 has an amplitude of $\pi F/2$:

$$\begin{aligned} &\Gamma_H(t) = F\{(\pi/4) + \sin\omega t - \tfrac{1}{3}\sin 3\omega t + \ldots\} \\ &\omega = 2\pi/T_0 \end{aligned} \tag{A4}$$

One can readily show, using standard methods, that

$$n_1(z) = sF\cosh\left(\frac{z}{L}\right)\left\{g\cosh\left(\frac{d}{L}\right) + \frac{D}{L}\sinh\left(\frac{d}{L}\right)\right\}^{-1} \tag{A5}$$

$$L = \sqrt{D/j\omega} = L_H(1-j) \tag{A6}$$

where $L_H = \sqrt{D/2\omega}$. Defining $\Delta G_1$ as the change of conductance at the fundamental frequency $\omega/2\pi$, we have from eqn. (4)

$$\Delta G_1 = -\frac{we\mu_p}{l}\,sFL^2\tanh\left(\frac{d}{L}\right)\left\{gL + D\tanh\left(\frac{d}{L}\right)\right\}^{-1} \tag{A7}$$

The above equation is complicated so that some simplification is desirable based on assumed conditions which must be satisfied experimentally. For $L_H \ll d$,

$$\omega \gg D/2d^2 \tag{A8}$$

From this condition, the high frequency limit of eqn. (A7) can be obtained:

$$-\Delta G_1 = \frac{we\mu_p sF}{l} \frac{2L_H}{g} \left\{1+j\left(1+\frac{\sqrt{2\omega D}}{g}\right)\right\}^{-1} \tag{A9}$$

The last term in the above equation is the phase factor

$$\tan\theta = 1+\sqrt{2\omega D}/g \tag{A10}$$

$\theta$ is the phase angle between the fundamental frequencies of the excitation function $\Gamma_H(t)$ and the change of conductance $\Delta G_1$. It can be extracted by Fourier series analysis from the measured change of conductance $\Delta G_1(t)$ at the fundamental frequency $\omega/2\pi$. From this type of measurement, the value of $\sqrt{\omega D}/g$ can be obtained.

For $L_H \gg d$, the angular velocity $\omega$ must be much smaller than $D/2d^2$ and the low frequency limit of eqn. (A7) can be obtained:

$$-\Delta G_1 = \frac{we\mu_p sFd}{lg} \left(1-\frac{j\omega d}{g}\right)^{-1} \tag{A11}$$

The last term of the above equation is the phase factor

$$\tan\gamma = -\omega d/g \tag{A12}$$

To satisfy the low frequency condition, $\omega$ would have to be so small that it would not be practical. The following numerical calculation shows this quite clearly.

$$T_0 = 2\pi/\omega \gg 4\pi d^2/D$$

$$D = 10^{-14}\ \text{cm}^2\ \text{sec}^{-1}$$

$$d = 1500\ \text{Å} = 1.5\times10^{-5}\ \text{cm}$$

$$4\pi d^2/D = 4\pi\times1.5^2\times10^{-10}/10^{-14} \approx 2.8\times10^5\ \text{sec} > 3\ \text{days}.$$

*Thin Solid Films*, *31* (1976) 39–51



# EPITAXIAL DEPOSITION OF SILICON CARBIDE FROM SILICON TETRACHLORIDE AND HEXANE

W. V. MUENCH AND I. PFAFFENEDER
*Institut A fuer Werkstoffkunde, Technische Universitaet, Hannover (Germany)*
(Received April 1, 1975; accepted June 17, 1975)

This paper describes the deposition of cubic and 6 H silicon carbide onto 6 H substrates. The influence of the deposition parameters (*e.g.* temperature, gas composition and flow rate) are discussed. Optimized conditions for the epitaxial growth of 6 H layers are deduced from these results. Similar deposition parameters can be used for the deposition of 15 R layers onto substrates of the same polytype. Doping experiments have been performed with nitrogen, ammonia, diborane and organometallic aluminum compounds.

The characterization of the epitaxial layers includes oxidation studies and electrolytic and molten KOH etching. The electrical properties of these layers are evaluated by the van der Pauw technique and by the differential capacitance method with evaporated Schottky contacts. Some preliminary results on electroluminescence are reported.

---

## 1. INTRODUCTION

Silicon carbide is considered a useful material for high temperature and optoelectronic semiconductor devices. The technology of silicon carbide, however, is extremely difficult and expensive. SiC crystals of marginal size can be produced by a sublimation process (the Lely technique) in the 2500°–2600 °C temperature range[1]. These crystals are rather imperfect and great difficulties are encountered in preparing material of sufficient purity. The diffusion of dopants in silicon carbide requires temperatures around 1900 °C, and special precautions are necessary to prevent the decomposition of silicon carbide at this temperature. The chemical vapor deposition of silicon carbide, on the other hand, can be performed at moderate expense and may even be less complicated than some of the techniques adopted for the vapor growth of III–V compounds.

Silicon carbide exists in several different modifications. These "polytypes" can be described in terms of different stacking orders of the Si–C double layers. The cubic (3 C) polytype corresponds to the zincblende lattice (stacking order ABCABC). There is also a hexagonal polytype (2 H) the lattice of which is identical with that of wurtzite (stacking order ABAB). Among the most abundant types of silicon carbide are the 6 H crystals with a unit cell of six Si–C layers (and hexagonal symmetry) and the 15 R crystals with a unit cell of fifteen Si–C layers (and trigonal symmetry). The band gap of silicon carbide increases with the percentage of "hexagonality", *i.e.* from 2.2 eV (3 C) to 3.2 eV (2 H).

An application of silicon carbide of major importance is in the fabrication of blue luminescent diodes. This application requires a band gap exceeding 2.8 eV. The 6 H polytype of silicon carbide ($W_G = 2.89$ eV) appears most suitable. The emphasis of this work, therefore, is on the deposition of 6 H layers onto substrates of the same polytype. It is obvious that an exact doping control is a prerequisite to the optimization of electroluminescent devices. Doping experiments were therefore performed with gaseous dopants ($N_2$, $NH_3$, $B_2H_6$) and organometallic aluminum compounds.

## 2. VAPOR GROWTH APPARATUS

The chemical vapor deposition of silicon carbide requires gaseous or liquid compounds containing silicon and carbon atoms. A variety of compounds, including $SiH_4$, $SiCl_4$, $CH_4$, $C_3H_8$, $C_3H_6$, $C_6H_{14}$, $CCl_4$ and $CH_3SiCl_3$, have been used by different research workers. Systems based on the decomposition of molecules comprising silicon and carbon atoms, such as $CH_3SiCl_3$, may be attractive from the point of view of simplicity; lack of flexibility (to change the Si:C ratio), however, is a serious disadvantage in this case. A system based on the reaction of silicon tetrachloride with hexane in a hydrogen ambient seems to be a reasonable compromise in terms of simplicity, flexibility and purity of the starting material.

A schematic representation of the growth system is shown in Fig. 1. Hydrogen gas, purified by a palladium cell, is metered and fed into the system through the bubblers containing $SiCl_4$ and $C_6H_{14}$, and through a bypass. The bubblers are kept at a common constant temperature ranging from $-20°$ to $-27$ °C. The total hydrogen flow rate is 3 l $h^{-1}$; this corresponds to a carrier gas velocity of approximately 10 cm $s^{-1}$ at the substrate surface.

The design of the water-cooled quartz reactor and the shape of the susceptor can be seen from Fig. 2. In the course of these investigations it was found important to place the substrate at a position with uniform flow rate; the vertical surface of the susceptor also provides a homogeneous temperature distribution. The cone-shaped top of the susceptor is necessary to give a smooth flow pattern. The susceptor temperature is measured pyrometrically before and during the deposition cycle. The output signal of a second pyrometer is fed into the control system of the r.f. generator. The reactor and the bubblers can easily be removed from the system for cleaning purposes (joints are not drawn in Fig. 1).

## 3. SUBSTRATE PREPARATION

P- and n-type crystals of the 6 H polytype produced by the Lely technique are carefully selected with the aid of the Laue diffraction pattern. The polarity of the basal planes is determined by thermal oxidation in wet oxygen. As shown previously[2], the oxidation rate of silicon carbide depends on the polytype, the

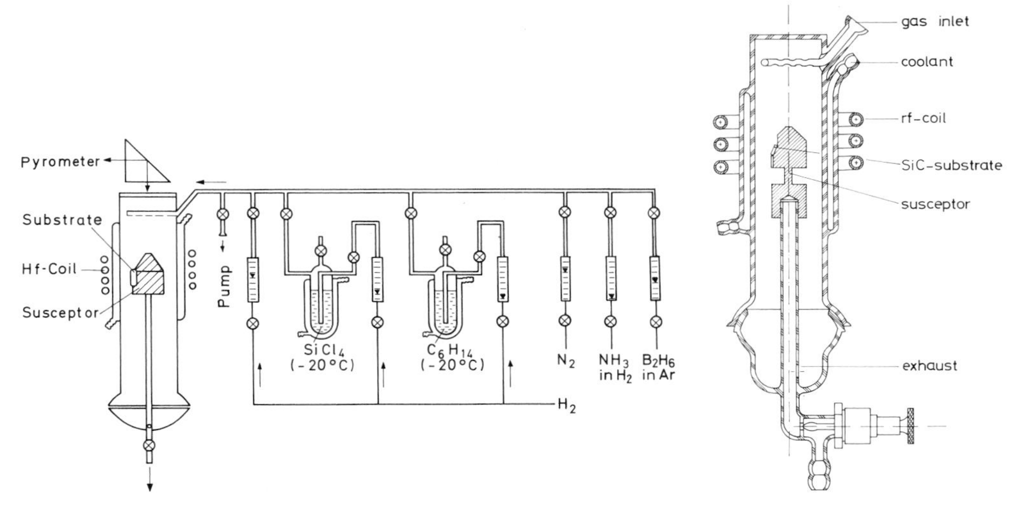


Fig. 1. Apparatus for chemical vapor growth of silicon carbide.
Fig. 2. Reactor tube with susceptor.

crystallographic orientation and the doping level (Table I). The (0001) basal plane (Si face, lowest oxidation rate) was used exclusively for the deposition experiments described in this paper.

Alternatively, the polarity discrimination may be performed by etching in molten potassium hydroxide (500 °C, 1 min). As seen from the micrographs (Fig. 3) the (0001) basal plane remains flat (with hexagonal etch pits), whereas the (000$\bar{1}$) face gets a dull appearance from etching in molten hydroxide.

The silicon face is carefully polished with boron carbide and wet diamond powder (grain size 15 μm, 3 μm and 1 μm). Surface cleaning is accomplished by organic solvents, hot nitric acid and hydrofluoric acid.

## 4. VAPOR GROWTH PROCESS AND GROWTH RATES

Each vapor growth experiment starts with a 4 min *in situ* etching process in pure hydrogen, removing a layer of approximately 1 μm. During this period, the flow rates through the bubblers are adjusted according to the deposition program. The deposition cycle is followed by a short (1 min) cooling period.

TABLE I

OXIDE THICKNESS AFTER OXIDATION IN WET OXYGEN AT 1070 °C FOR 6 h

| | *6 H n-type* | | *6 H p-type* | *15 R n-type* | *15 R p-type* |
|---|---|---|---|---|---|
| | *heavily doped* | *lightly doped* | | | |
| (000$\bar{1}$) face (Å) | 2730 | 2560 | 2420 | 2810 | 2200 |
| (0001) face (Å) | 310 | 300 | 450 | 350 | 420 |

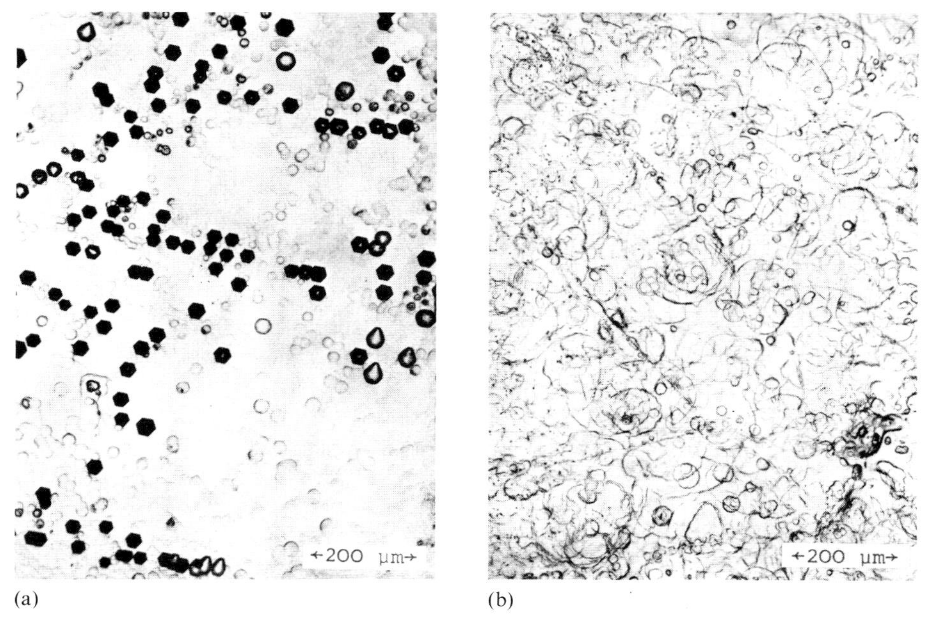


(a) (b)

Fig. 3. SiC after etching in molten KOH showing (a) the (0001) face and (b) the $(000\bar{1})$ face.

The influence of the following experimental parameters was studied during the early stages of the epitaxy program: deposition temperature, gas composition and Si:C ratio. Table II gives a survey on the range of growth conditions.

As expected, there is a nearly linear relationship between the growth rate and the $SiCl_4$ concentration (with the deposition temperature and Si:C ratio fixed). At 1760 °C the deposition rate increases by a factor of three if the hexane flow is kept at $2.8 \times 10^{-5}$ mol min$^{-1}$ and the silicon to carbon ratio is changed from 1:3 to 1:1. A 50% decrease is noticed when the deposition temperature is raised from 1700° to 1850 °C.

## 5. POLYTYPISM OF VAPOR-GROWN SiC LAYERS

Cubic (3 C) and hexagonal (6 H) silicon carbide layers were deposited onto 6 H substrates. The typical surface morphology of cubic SiC layers is shown in Fig. 4, a short etching in molten KOH revealing trigonal structures. The surface of 6 H layers, on the other hand, is smooth or slightly wavy, the morphology probably being related to the gas flow pattern. Figure 5 is a photomicrograph of the surface of a 6 H layer (with an ohmic contact).

Some more information about the polytypism of vapor-grown layers is obtained by angle lapping and subsequent oxidation in wet oxygen at 1070 °C for 6 h. With these oxidation conditions a relatively thick oxide layer is produced on the bevelled surface of the cubic material. This oxide layer appears blue under illumination with an incandescent lamp. Corresponding oxide layers grown on 6 H material are much thinner and exhibit a light-brown interference coloring.

TABLE II

RANGE OF EXPERIMENTAL PARAMETERS FOR DEPOSITION OF SILICON CARBIDE

| | *Minimum* | *Maximum* |
|---|---|---|
| Deposition temperature (°C) | 1700 | 1850 |
| Temperature of bubblers (°C) | −27 | −20 |
| Hydrogen flow rate through $SiCl_4$ (mol $min^{-1}$) | $0.4 \times 10^{-3}$ | $2.4 \times 10^{-3}$ |
| Silicon to carbon ratio | 1:3.5 | 1:1 |
| Deposition time (min) | 30 | 180 |

Fig. 4. Surface morphology of cubic epitaxial layers after etching in molten KOH (500 °C, 1 min).

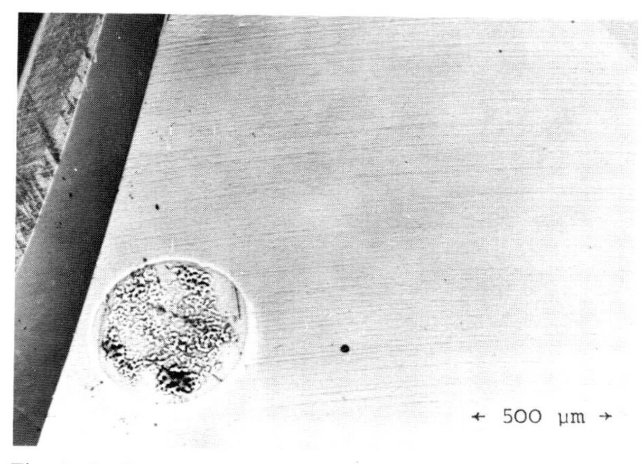


Fig. 5. Surface structure of a 6 H epitaxial layer (ohmic contact at the bottom left).

The boundaries between cubic and hexagonal material are therefore clearly seen by microscope inspection.

It has been found previously that a high deposition temperature and a low growth rate favor the generation of the hexagonal polytype of silicon carbide[3]. The results of the present investigation indicate, however, that there is not an abrupt transition from one deposition condition with cubic growth to another with hexagonal growth. Figure 6 is a photomicrograph of an oxidized angle lap of a cubic layer grown at a high deposition rate. The typical surface morphology of cubic material is seen at the top. It should be noted that the transition from hexagonal to cubic material is not located at the interface between the substrate and the epitaxial layer, but occurs along some planes which are obviously related to certain crystallographic orientations.

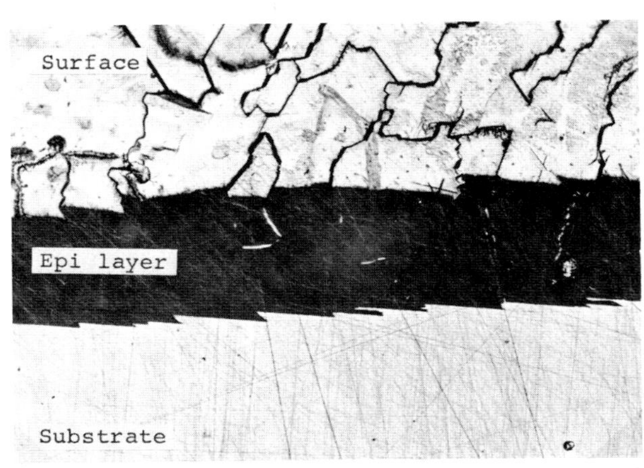


Fig. 6. Oxidized angle lap of a cubic epitaxial layer on a 6 H substrate.

A slight reduction of the growth rate yields an epitaxial layer in which the cubic polytype is still dominant; isolated 6 H regions of limited size sometimes propagate to the surface (Fig. 7). A further reduction of the growth rate gives rise to the deposition of 6 H layers with minor cubic inclusions (Fig. 8).

Layers of the 6 H polytype are routinely produced with the following deposition conditions:

growth temperature 1850 °C
bubbler temperature −20 °C
vapor pressure of silicon tetrachloride 27 Torr
vapor pressure of hexane 13 Torr
hydrogen flow through silicon tetrachloride 0.010 l $min^{-1}$
hydrogen flow through hexane 0.012 l $min^{-1}$
silicon tetrachloride flow rate $1.5 \times 10^{-5}$ mol $min^{-1}$
hexane flow rate $0.87 \times 10^{-5}$ mol $min^{-1}$
total hydrogen flow rate 3 l $min^{-1}$
growth time 100 min
growth rate 0.06–0.12 μm $min^{-1}$

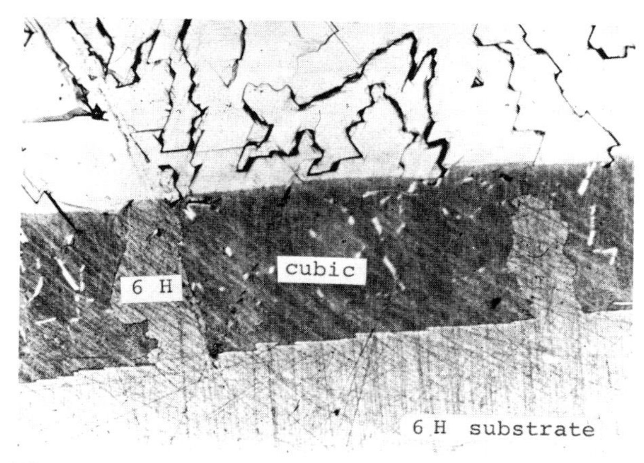


Fig. 7. Oxidized angle lap of an epitaxial layer, cubic material dominating.

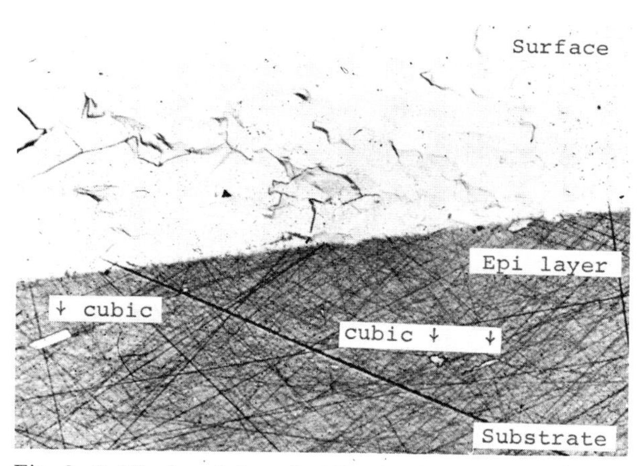


Fig. 8. Oxidized angle lap of a 6 H epitaxial layer with cubic inclusions.

The growth rate depends on the actual surface temperature and on the gas flow pattern. Consequently there is a range of epitaxial layer thickness due to variations in substrate thickness and substrate shape. Cubic inclusions are often found near the edges of the substrate.

These deposition conditions may also serve for the epitaxial growth of 15 R layers on 15 R substrates. The polytype of the substrate and the epitaxial layer can be conveniently revealed by molten KOH etching. Figure 9(a) is a photomicrograph of etch pits in 6 H silicon carbide while Fig. 9(b) shows typical etch pits of the 15 R polytype. Small etch pits which are entirely within the epitaxial layer and large etch pits which penetrate into the substrate exhibit the same shape. It is therefore concluded that the above-mentioned growth conditions are able to generate both 6 H and 15 R layers, the polytype depending on the crystal structure of the substrate.

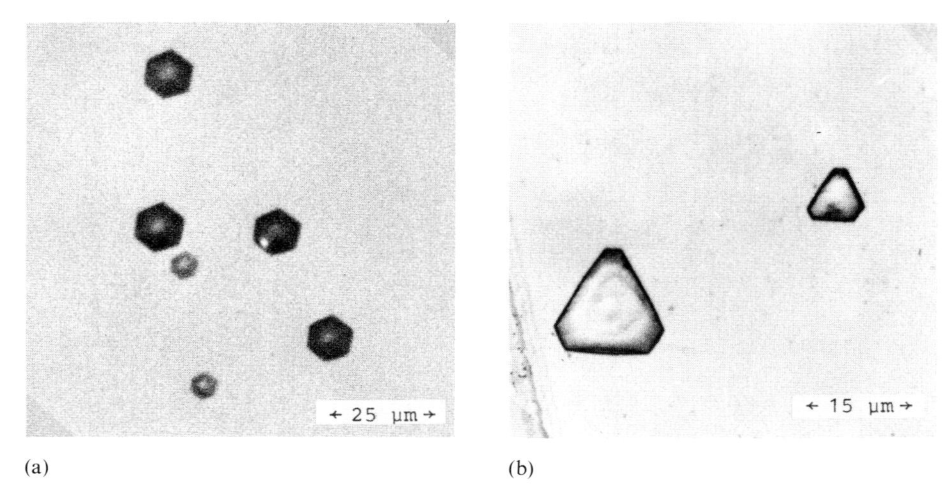


Fig. 9. Etch pits (a) in a hexagonal (6 H) layer on a 6 H substrate and (b) in a 15 R layer on a 15 R substrate.

Cubic growth not only coexists with hexagonal growth (as shown in Figs. 6–8) but can also occur simultaneously with 15 R growth at moderate deposition rates. The etch-pit orientation of the cubic regions is correlated with that of the 15 R regions (Fig. 10).

Cubic layers are not, in general, single crystal. Grain boundaries can be seen on the top surface (Fig. 4) and on the angle-lapped oxidized faces (Fig. 7). Electron diffraction studies have revealed a twin formation within cubic silicon carbide layers.

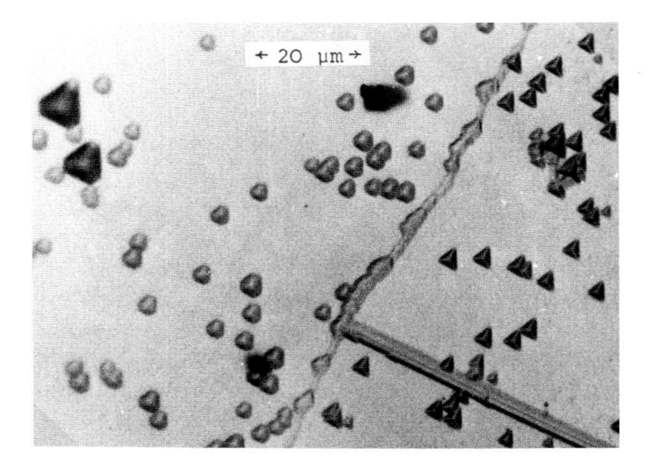


Fig. 10. Etch pits in an epitaxial layer on a 15 R substrate: 15 R polytype on the left-hand side, cubic material on the right.

## 6. ELECTRICAL CHARACTERIZATION OF EPITAXIAL LAYERS

The conductivity type of epitaxial silicon carbide layers is determined by standard metal point probe techniques. Most p–n junctions can be revealed by the above-mentioned oxidation method due to small differences in the oxidation rates, as indicated in Table I.

Electrolytic etching in a 2:1 mixture of alcohol and 40% HF is used for the delineation of $p^+$–p and $n^+$–n junctions and for those p–n junctions in which both sides are very lightly doped. The delineation, which is accomplished partly by surface attack and partly by staining, can be markedly enhanced by a subsequent oxidation in wet oxygen. Thus the combination of electrolytic etching and oxidation has been found to be successful for the delineation of all junctions of practical interest. Furthermore, the existence of doping inhomogeneities can be revealed by this technique. Figure 11 shows an angle lap of an n-type epitaxial SiC layer on an n-type substrate. It is clearly seen that there are impurity striations in the substrate whereas the doping of the epitaxial layer is uniform.

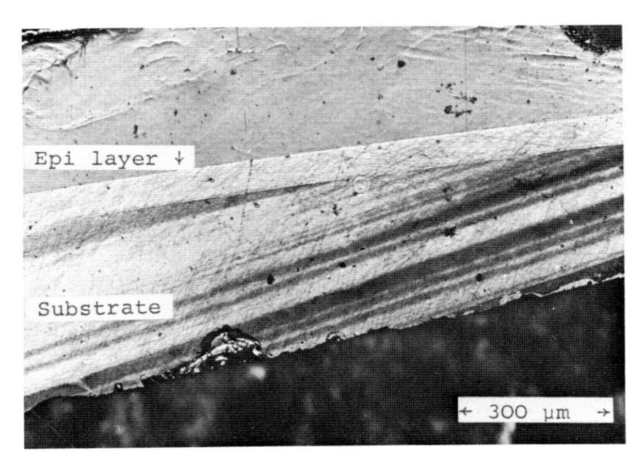


Fig. 11. Angle lap of an n-type epitaxial layer on an n-type substrate (etched electrolytically prior to oxidation).

The point-contact breakdown voltage gives a quick check on the doping level of the epitaxial layers. A calibration curve relating the point-contact breakdown voltage to the carrier concentration is shown in Fig. 12.

More detailed studies of the electrical properties of silicon carbide layers can be performed with the van der Pauw technique or with the differential capacitance method. The realization of very shallow non-rectifying contacts is a major problem in the characterization of silicon carbide and in the technology of SiC devices.

Ohmic contacts to p-type epitaxial layers may be obtained by alloying Al/Si to silicon carbide at 1300 °C. Alternatively, an electron-beam-evaporated tungsten layer may be alloyed to silicon carbide at 1800 °C. The latter type of contact, however, is recommended only for layers with a carrier concentration exceeding $10^{17}$ $cm^{-3}$.

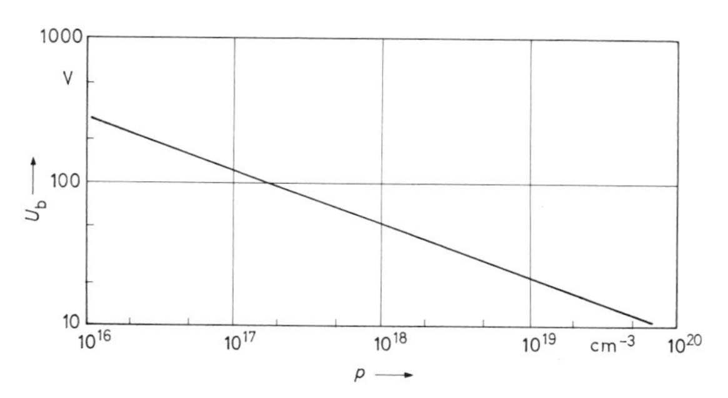


Fig. 12. Relation between the point-contact breakdown voltage and the carrier concentration in p-type silicon carbide.

The most reliable ohmic contacts to thin n-type silicon carbide layers are obtained by sintering a nickel film at 1100 °C for 5 min. A gold topping is required for thermocompression bonding. The contact area must be carefully checked since nickel tends to diffuse deeply in silicon carbide along grain boundaries or other gross imperfections.

The differential capacitance method is a useful tool for the evaluation of p and n layers on substrates of the same conductivity type. Schottky contacts are produced by evaporation of chromium (~400 Å) and gold (~1000 Å) through a metal mask. Alternatively, the area definition can be accomplished by standard photolithographic techniques. The *I–V* characteristic of a Cr/Au contact on p-type silicon carbide is shown in Fig. 13. These contacts are stable up to 150 °C.

Figure 14(a) shows an example of the $1/C^2$ plot for three Schottky diodes on a p-type silicon carbide epitaxial layer. A range of hole concentrations between $5 \times 10^{16}$ and $6 \times 10^{16}$ $cm^{-3}$ is deduced from the slopes. In thin and lightly doped epitaxial layers the space charge region of the Schottky contact extends to the substrate. The initial slope (Fig. 14(b)) then yields the carrier concentration whilst the thickness of the epitaxial layer can be calculated from the saturation value of

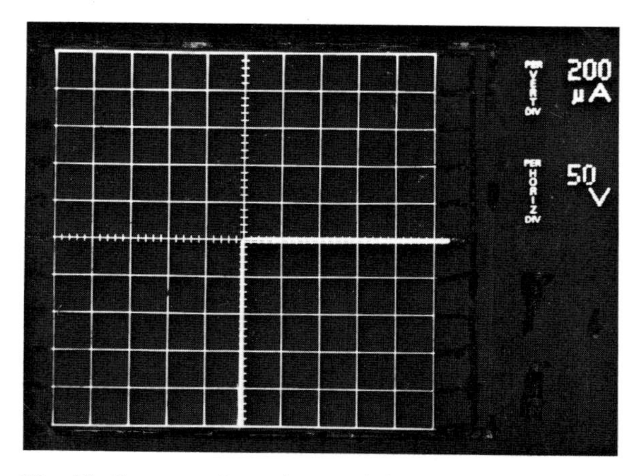


Fig. 13. Current–voltage characteristic of a Cr/Au Schottky contact on p-type silicon carbide.

$1/C^2$. The example of Fig. 14(b) gives an electron concentration of $3 \times 10^{15}$ cm$^{-3}$ and a thickness range of 3.6–4.4 μm.

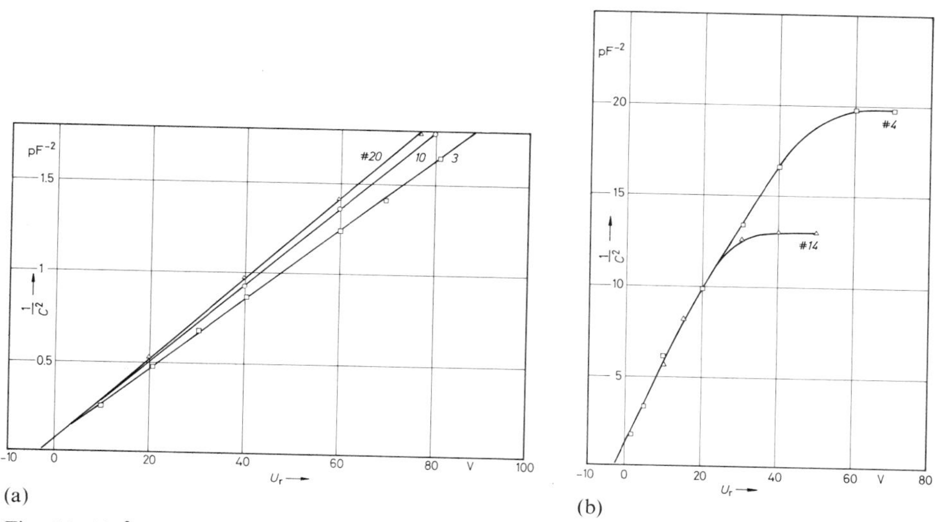


Fig. 14. $1/C^2$ *vs.* reverse voltage of Schottky contacts on (a) a thick p-type epitaxial layer and (b) a thin n-type epitaxial layer.

## 7. DOPING EXPERIMENTS

Doping of epitaxial layers is most easily accomplished by adding gaseous compounds, such as $B_2H_6$, $NH_3$ or $N_2$, to the $SiCl_4/C_6H_{14}/H_2$ mixture. As seen in Fig. 15, the most effective doping is obtained with diborane. However, the same partial pressure of diborane would give a hole concentration in epitaxial silicon about one order of magnitude higher than that in silicon carbide (due to the incomplete ionization of acceptors in SiC at room temperature).

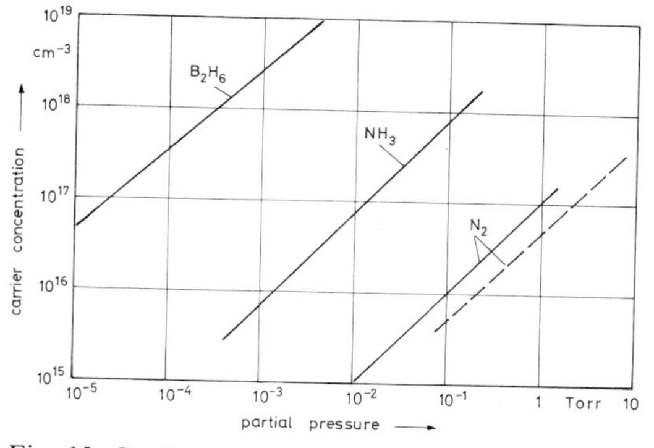


Fig. 15. Carrier concentration in epitaxial silicon carbide *vs.* dopant partial pressure.

The lowest doping efficiency is found when nitrogen gas is added. This is obviously due to the high dissociation energy of the $N_2$ molecule. The results of two sets of experiments with $N_2$ doping are shown in Fig. 15. The solid line was obtained by the differential capacitance technique (n layers on $n^+$ substrate), whereas the broken line results from van der Pauw measurements (n layers on p substrates). Autodoping effects may be at least partly responsible for the discrepancies.

Higher donor concentrations can be achieved when ammonia gas is introduced. Extreme care has to be exercised, however, to avoid moisture, which catalyzes a reaction with $SiCl_4$ and the precipitation of $NH_4Cl$.

Preliminary experiments indicate that aluminum doping is possible when suitable organometallic aluminum compounds are added to the hexane. A 1% triethyl aluminum solution yields a hole concentration of about $2\times10^{15}$ $cm^{-3}$ (vapor pressure of triethyl aluminum $4.3\times10^{-3}$ Torr at $-20\,°C$). With the same percentage of trimethyl aluminum (vapor pressure 0.6 Torr at $-20\,°C$) layers with a hole density of $5\times10^{16}$ $cm^{-3}$ have been obtained.

## 8. ELECTROLUMINESCENCE

Previous investigations have shown that light emission in the blue range of the spectrum is generated in 6 H silicon carbide by recombination via nitrogen or aluminum impurity levels[4]. Consequently, the design of blue electroluminescent diodes is based on hole injection into nitrogen-doped silicon carbide or on electron injection into aluminum-doped material.

Double-epitaxial diodes prepared by silicon carbide vapor growth have given the best quantum yield so far. Starting with a p-type substrate ($p\approx5\times10^{16}$ $cm^{-3}$), a lightly doped p layer is deposited first. The top layer is doped with nitrogen (by addition of $N_2$ or $NH_3$). Figure 16 is the photomicrograph of an angle-lapped double-epitaxial structure, with junction delineation by electrolytic etching and subsequent oxidation, as described earlier.

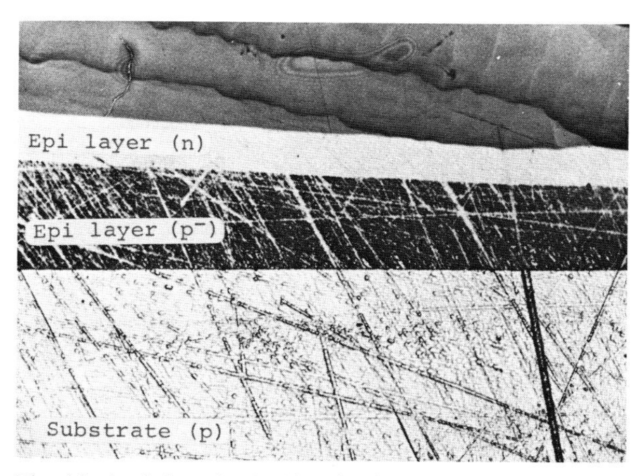


Fig. 16. Angle lap of a double-epitaxial ($np^-p$) structure.

Contacts to the substrate are made by alloying Al/Si; the top contact is sintered nickel. Figure 17 shows the emission spectra of two vapor-grown double-epitaxial SiC diodes. The external quantum yield is about $5 \times 10^{-6}$.

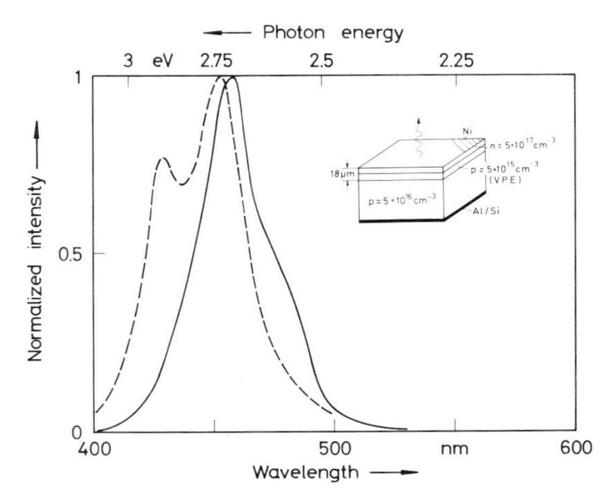


Fig. 17. Emission spectrum of double-epitaxial SiC diodes.

## 9. CONCLUSION

6 H silicon carbide can be conveniently deposited onto substrates of the same polytype by a reaction of silicon tetrachloride and hexane, with hydrogen as the carrier gas. For (0001) faces the following growth conditions have been found to be most suitable:

growth temperature 1850 °C
silicon tetrachloride flow rate $1.5 \times 10^{-5}$ mol min$^{-1}$
hexane flow rate $0.87 \times 10^{-5}$ mol min$^{-1}$
hydrogen flow rate 3 l min$^{-1}$

The growth rate is in the 0.06–0.12 µm min$^{-1}$ range.

Nitrogen gas doping yields electron concentrations from $10^{15}$ to $10^{17}$ cm$^{-3}$, approximately. A substantially higher doping level is obtained with ammonia gas. Doping with diborane is a convenient method of producing p-type layers.

## REFERENCES


1 W. F. Knippenberg, *Philips Res. Rep.*, *18* (1963) 161.
2 W. v. Muench and I. Pfaffeneder, *J. Electrochem. Soc.*, *122* (1975) 642.
3 W. Spielmann, *Z. Angew. Phys.*, *19* (1965) 93.
4 J. M. Blank, *Proc. Int. Conf. on Silicon Carbide, Miami Beach, 1973*, Univ. of South Carolina Press, Columbia, S.C., p. 550.

*Thin Solid Films*, *31* (1976) 53–67



# ELECTRON MICROSCOPE STUDY OF EPITAXIAL SILICON FILMS ON SAPPHIRE AND DIAMOND SUBSTRATES

A. G. CULLIS*
*Bell Laboratories, Murray Hill, N. J. 07974 (U.S.A.)*
G. R. BOOKER
*Department of Metallurgy, University of Oxford, Oxford (Gt. Britain)*
(Received March 18, 1975; accepted June 17, 1975)

In the present work, Si films were deposited upon both (0001) sapphire and (111) diamond substrates by sublimation or evaporation in UHV. The main aim was to study details of the film structures, especially at early stages of the growth process, using direct electron microscope methods.

Initial growth in both systems was by nucleation, leading to the formation of three-dimensional growth centers which eventually overlapped to give a continuous film. For Si deposited on sapphire, the smallest growth centers (~ 100 Å) were directly observed to be composed of one or other of two rotational twins, each exhibiting a relationship of the type $(111)_{Si} \| (0001)_{Al_2O_3}$ and $[1\bar{1}0]_{Si} \| [11\bar{2}0]_{Al_2O_3}$. No definite evidence was obtained for the production of large-scale networks of misfit dislocations at the Si–sapphire heterojunction, although the results suggested that such dislocations may have been present in localized regions. Mottled background contrast features observed in TEM images were interpreted in terms of a probable reaction between the Si and the sapphire during film growth. Reaction products (*e.g.* $Al_2SiO_5$) remaining at the interface were thought to be associated with film defect production. The thicker Si films contained many defects including stacking faults and microtwins, and the thickest film studied (2.9 μm) contained relatively large blocks of misoriented Si.

For Si deposited on diamond, the thicker films showed evidence for partial retention of the original (111) substrate orientation. However, the films were highly defective, containing microtwins and regions of Si exhibiting general crystallographic rotations.

## 1. INTRODUCTION

A detailed investigation has been made of Si films grown under ultrahigh vacuum (UHV) conditions upon heated single-crystal sapphire substrates. Some results have also been obtained for Si films grown upon diamond substrates. Technological interest in such film/substrate combinations arises because

* Present address: Royal Radar Establishment, Malvern, Gt. Britain.

the electrically insulating nature of the substrates offers advantages for the fabrication of certain transistor devices.

Previous work on the Si/sapphire system (see reviews by Filby and Nielsen[1], Cullen[2], Joyce[3] and Manasevit[4]) has been mainly concerned with the growth of Si films by either vapor decomposition[5–8] or vacuum evaporation[8–13] methods. Such grown films were not generally uniform single-crystal and contained numerous defects including microtwins and stacking faults. Films often possessed poor electrical carrier mobilities when thin, but the values improved as film thickness increased[14] suggesting the presence of defects near the Si–substrate interface. This is in agreement with $He^+$ ion backscattering studies[15] and conductance measurements under pulsed electron bombardment[16]. However, little detailed structural information has been obtained concerning the interface region, mainly due to the difficulty of thinning the sapphire for direct transmission electron microscope (TEM) examinations of films attached to their substrates. For thick Si films, TEM studies[17] have shown that the number density of dislocations decreases with distance from the Si–sapphire interface. Regarding Si films grown on single-crystal diamond, as yet few structural examinations seem to have been made.

The main aim of the present work was to make detailed structural examinations, using electron microscope methods, of Si films deposited in UHV upon single-crystal sapphire (and diamond) substrates for a variety of growth conditions. Special attention was given to the crystallographic orientation and perfection of the films, and to the nature of the interface region. The sapphire substrate surface orientation was (0001), this being the growth habit of the substrate platelets used in this study. Recently, the electrical properties of (111) Si grown by vapor decomposition on (0001) sapphire were shown to be comparable with or slightly better than those of (001) Si grown on $(1\bar{1}02)$ sapphire[18], the epitaxial combination commercially favored at present.

## 2. EXPERIMENTAL PROCEDURE

The apparatus and general procedures used for the Si deposition were similar to those described by the authors[19] for the epitaxial growth of Si films on Si substrates. Briefly, the depositions were performed in a stainless-steel apparatus which attained a base pressure of about $10^{-10}$ torr, this deteriorating to about $10^{-8}$ torr during experimental runs. The deposition was by sublimation or evaporation from a disc of Wacker, float-zoned, 300 Ω cm Si which was electron-beam heated from the back face.

Sapphire substrates were (0001) platelets of high structural perfection that had been grown by a vapor transport method. Individual platelets were typically 10 mm across, within the thickness range 0.5–3 μm, and their surfaces were step-free to the limits of resolution of a Tolansky multiple-beam optical interferometer. Deposition was made onto the as-grown (0001) faces. The diamond substrates were prepared by cleaving type IIA crystals and the resulting pieces were (111) plates exhibiting fine cleavage steps, being several millimeters across and typically of 0.5 mm thickness. These pieces were cleaned with boiling HCl, followed by an $HNO_3$, HF, $CH_3CO_2H$ mixture (CP4) and then methanol, and Si deposition was made onto the (111) cleaved faces.

Each sapphire or diamond substrate was heated in UHV, being placed either directly upon a gas-polished resistance-heated Si bar, or in a recess made in the upper surface of such a bar. The substrate temperature was taken to be the temperature of the Si bar and was determined using an optical pyrometer. Although this may have resulted in an error of a few tens of degrees Centigrade, it was nevertheless adequate for the present work. Deposition rates were determined from optical interferometer measurements of the thickness of Si deposited onto a cooled glass slide placed adjacent to the substrate heating bar. A shutter was used to regulate Si deposition.

Methods of specimen examination included use of the optical microscope, the TEM employing direct transmission through the specimens, and the scanning electron microscope (SEM). The specimens with sapphire substrates were prepared for TEM examination by ion-beam thinning from the substrate side. Specimens with diamond substrates were mechanically ground and polished from the substrate side with diamond powders to a thickness of about 25 μm, after which they were ion-beam thinned to achieve electron transparency. In some instances, a 30 Å carbon film was evaporated onto the substrate side after the thinning to eliminate electrical charging effects during the TEM studies. Some of the diamond substrate specimens were examined using a TEM with an electron accelerating voltage of 1 MV, rather than the conventional 100 kV.

## 3. RESULTS

### *3.1. Silicon on sapphire*

When Si films were deposited at a rate of 0.5 Å $s^{-1}$ onto substrates held at 900 °C, for a mean film thickness of only 220 Å, TEM studies showed that separate Si growth centers were present. The centers had lateral dimensions ranging upwards from around 100 Å; a typical area is shown in Fig. 1. The larger centers had a number density in the region of $10^9$ $cm^{-2}$, and the substrate surface both between and beneath the centers exhibited fine spotted or mottled contrast features.

The interface region between the Si and the sapphire was carefully studied to investigate the possibility that arrays of misfit dislocations might have been present. However, only weak TEM image features were generally observed although, for certain diffraction conditions, these took the form of periodic contrasts with approximate spacings from less than 30 Å to about 70 Å. Fringes of about 40 Å spacing are weakly visible in Fig. 1, and an additional example is given in Fig. 2. Fringes which gave the strongest image contrast were usually found in localized areas, and they were often irregular in form (see Fig. 1).

Selected area electron diffraction (SAED) patterns were obtained from areas similar to that of Fig. 1, and the patterns contained many spots produced by double diffraction effects. Figure 3 gives an example where the primary diffraction spots are labelled and this is sufficient to confirm directly the epitaxial relationship $(111)_{Si}||(0001)_{Al_2O_3}$ and $[1\bar{1}0]_{Si}||[11\bar{2}0]_{Al_2O_3}$.

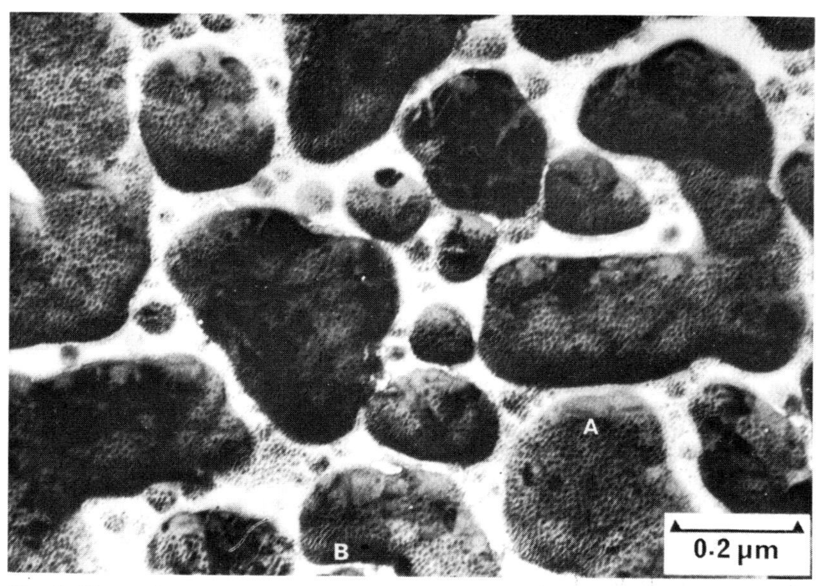


Fig. 1. Si growth centers formed at a deposition rate of 0.5 Å $s^{-1}$ with a sapphire substrate temperature of 900 °C. Fringe contrast is weakly visible, for example, near A, while more prominent but less regular periodic contrasts occur near B. Note the mottled background features. (Transmission electron micrograph, dark field using both Si $2\bar{2}0$- and sapphire $11\bar{2}0$-type reflections.)

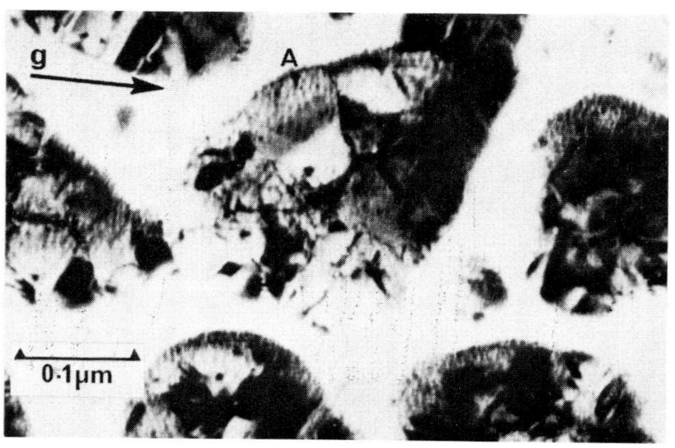


Fig. 2. Si growth centers formed as in Fig. 1 and showing fringe contrast, for example, near A. (Transmission electron micrograph, dark field; conditions as for Fig. 1.)

Further TEM studies revealed that the Si growth centers were composed of essentially equal proportions of two 60° rotational twins, each with $[1\bar{1}0]_{Si} \| [11\bar{2}0]_{Al_2O_3}$. In order to distinguish these twins in the microscope it was necessary to employ Bragg reflections with $\boldsymbol{g}$ vectors not parallel to the foil plane, so that rotationally displaced inclined Si planes were involved. Accordingly, Fig. 4(a) shows a pair of bright field micrographs of the same specimen area taken using

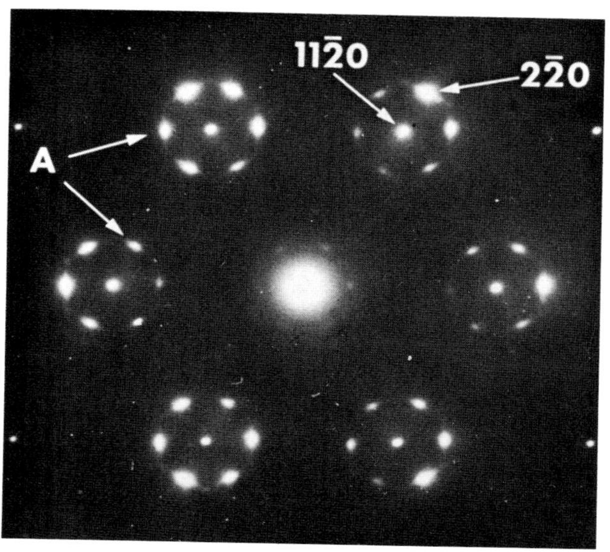


Fig. 3. Typical SAED pattern, showing both Si ($2\bar{2}0$-type) and sapphire ($11\bar{2}0$-type) spots, together with others of the type labelled A produced by double diffraction.

adjacent, 60° displaced, $11\bar{1}$-type twin reflections. Figure 4(b) is a diagram which illustrates the manner in which the usual Si single-crystal Kikuchi bands around a [111] diffraction pole are duplicated in the presence of 60° rotational twins. The various regions composed of the two different twins in Fig. 4(a) can be clearly seen, and the dark regions in each case are those which actively diffracted intensity out of the main electron beam. Detailed observations showed that the smallest centers were composed exclusively of either one or other of the two twins. Also, although the larger growth centers, formed by coalescence, did not appear to exhibit pronounced geometrical shapes, where they were bounded by straight edges these were sometimes parallel to $<1\bar{1}0>$ directions in the Si.

Si films were also grown at 900 °C with a deposition rate of 20 Å $s^{-1}$. For a mean deposit thickness of about 300 Å, initial growth centers had already apparently overlapped to give a film with a fairly fine channelled structure (Fig. 5). Substantial twinning had taken place and the film contained many defects. The Si–sapphire interfacial region gave mottled background contrast, although this was weaker than the corresponding contrast given by the films grown at only 0.5 Å $s^{-1}$. Also, SAED patterns showed that there was a small amount of film misorientation ($\sim \pm 5°$) about the preferred relationship.

When a Si film was grown at the higher rate up to a mean thickness of about 0.15 μm, few channels remained in the deposit and it achieved more uniform coverage of the substrate (Fig. 6). Diffraction information once again showed that there were small crystal rotations about the preferred orientation, and the linear features evident in the micrograph were probably given by microtwin lamellae (but see below).

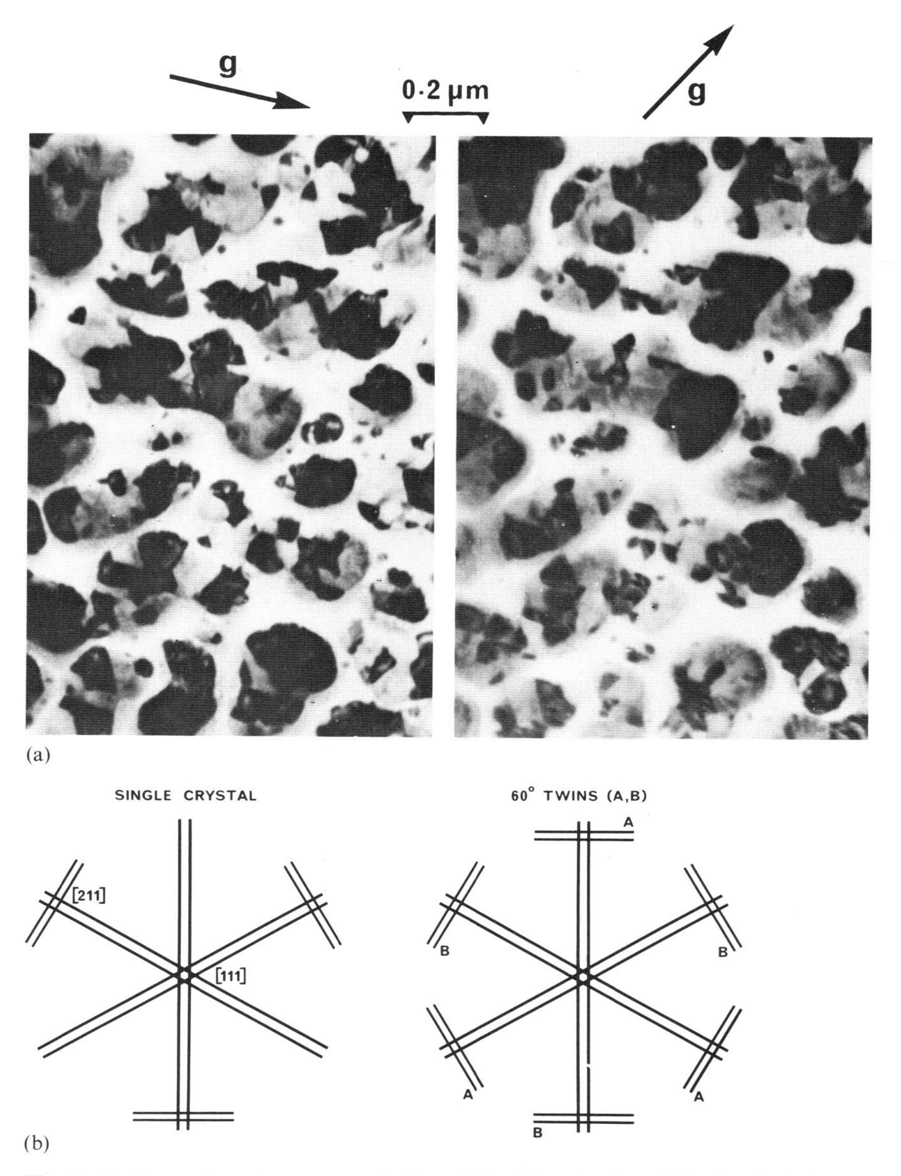


Fig. 4. (a) Si growth centers composed of two 60° rotational twins which give dissimilar contrast in the two images. (Transmission electron micrographs, same specimen area, bright field, with different operating $11\bar{1}$-type twin reflections.) (b) Kikuchi band diagrams around the [111] pole, illustrating the relationship between the transmission electron diffraction patterns given by 60° rotational Si twins A and B.

For comparison with these studies, Si films were grown at 1100 °C and at a rate of 15 Å $s^{-1}$. At a mean thickness of about 600 Å the film covered most of the substrate and only relatively small channels remained. Nevertheless,

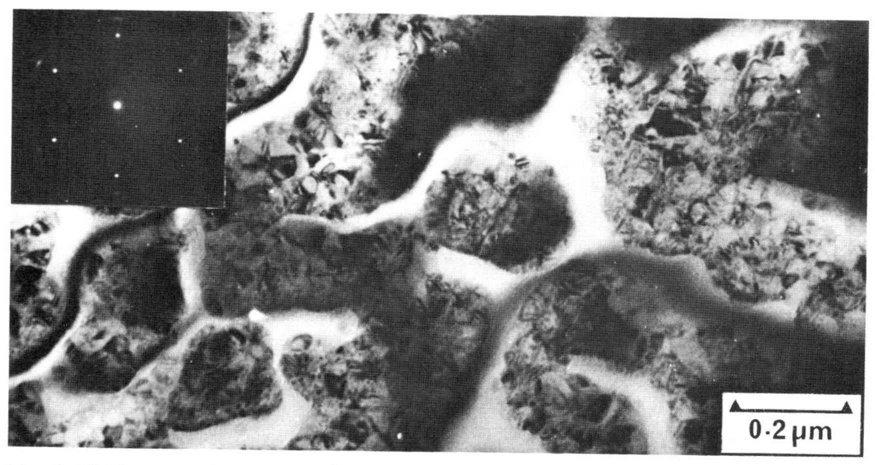


Fig. 5. Si film approximately 300 Å in thickness grown at 20 Å $s^{-1}$ with a sapphire substrate temperature of 900 °C. (Transmission electron micrograph, bright field; SAED pattern inset.)

Fig. 6. Si film approximately 0.15 μm in thickness grown at 20 Å $s^{-1}$ with a substrate temperature of 900 °C. (Transmission electron micrograph, bright field; SAED pattern inset.)

as shown in Fig. 7, the film contained an extremely high density of planar defects exhibiting fringe contrast. These were probably both stacking faults and microtwin lamellae, although it was not always a simple matter to distinguish between them since they would often be expected to give similar image contrast[20]. Bright field micrographs also indicated that the Si–sapphire interface region yielded generally mottled contrast of the type described previously. Furthermore, SAED patterns showed that the Si was present almost exclusively with the preferred epitaxial relationship.

When Si was deposited under the conditions just described up to a total thickness of about 2.9 μm, film uniformity was very good. This is exemplified

Fig. 7. Si film approximately 600 Å in thickness grown at 15 Å $s^{-1}$ with a substrate temperature of 1100 °C. (Transmission electron micrograph, dark field, using the Si $2\bar{2}0$-type reflection indicated in the SAED pattern.)

in Fig. 8, which shows an SEM image of a cleaved cross section of a film grown upon a substrate of almost identical thickness. When the outer region of such a Si film was examined in the TEM it was found to contain quite a high density of defects, including relatively large blocks of misoriented Si. Figure 9 gives both bright and dark field micrographs of the same specimen area, showing the defects present in the film, including some microtwins. Occasional narrow channels still penetrated through the Si, although it is not certain whether they extended down to the substrate surface.

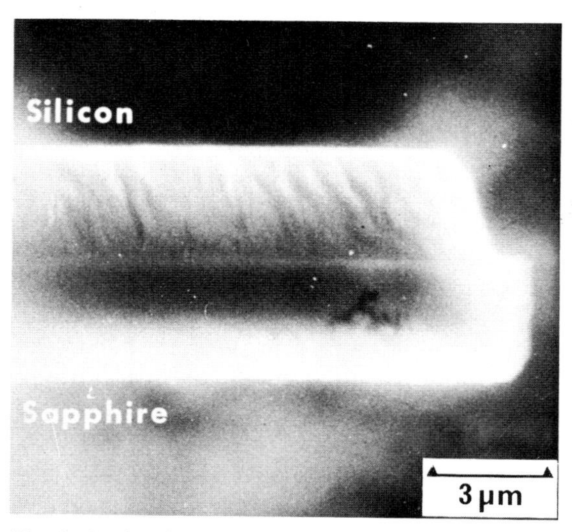


Fig. 8. Section through a Si film approximately 2.9 µm in thickness, grown at 15 Å $s^{-1}$ with a sapphire substrate temperature of 1100 °C. (Scanning electron micrograph, secondary emission mode.)

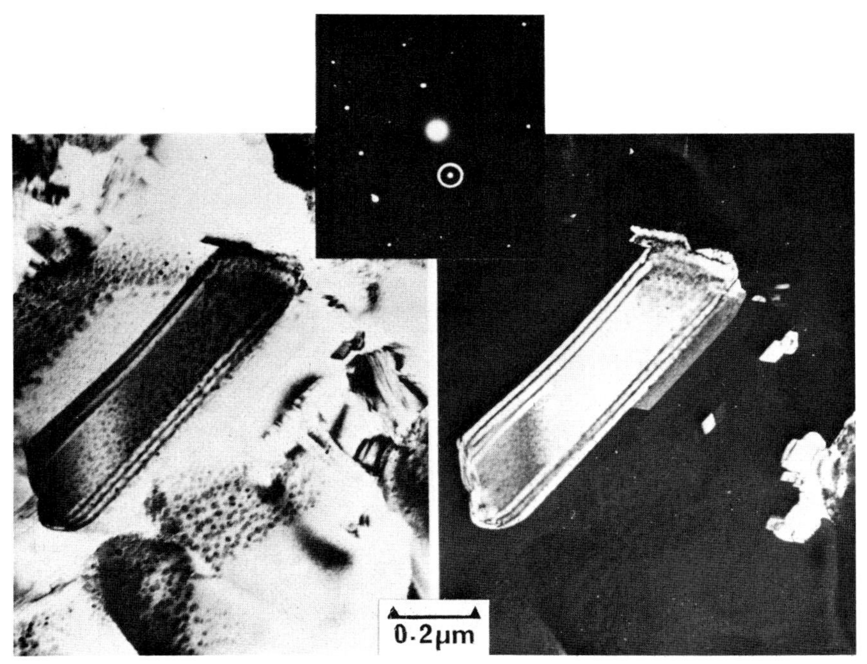


Fig. 9. Similar film to that shown in Fig. 8, illustrating the general defective nature and the presence of misoriented Si. (Transmission electron micrographs, bright field (left) and dark field (right); SAED pattern inset.)

### *3.2. Silicon on diamond*

For a substrate temperature of 900 °C, when Si was deposited onto (111) diamond at a rate of 1 Å $s^{-1}$ up to a mean film thickness of about 160 Å, subsequent examination in the SEM showed that separate small growth centers were present in a number density of approximately $3 \times 10^9$ $cm^{-2}$ (see Fig. 10). The micrograph taken with 0° of specimen tilt illustrates that the centers were sometimes approximately triangular in shape.

When the Si deposition at 1 Å $s^{-1}$ was continued up to a mean film thickness of about 0.4 μm the film became continuous. As shown in Fig. 11, examination in the SEM indicated that the surface of the film was somewhat irregular, having a rough appearance. General information about the crystallographic structure of the film was deduced from electron channelling patterns[21] obtained in the SEM from the film surface. Such a pattern is shown in Fig. 11 and it exhibits a diffuse [111] pole. This indicates that the film had an overall preferred orientation identical to that of the substrate, *i.e.* $(111)_{Si} \| (111)_{diamond}$. The diffuseness of the bands further indicates that the film contained a very high density of crystallographic defects, in accordance with studies[22] of channelling pattern contrast.

Some studies of the thicker Si films have been carried out in the 1 MeV TEM. Figure 12(a) illustrates the rippled structure near the film surface, the image having a channelled appearance. The film contained a very large number of defects, including stacking faults, microtwins and Si with general misorientations (Fig. 12(b)).

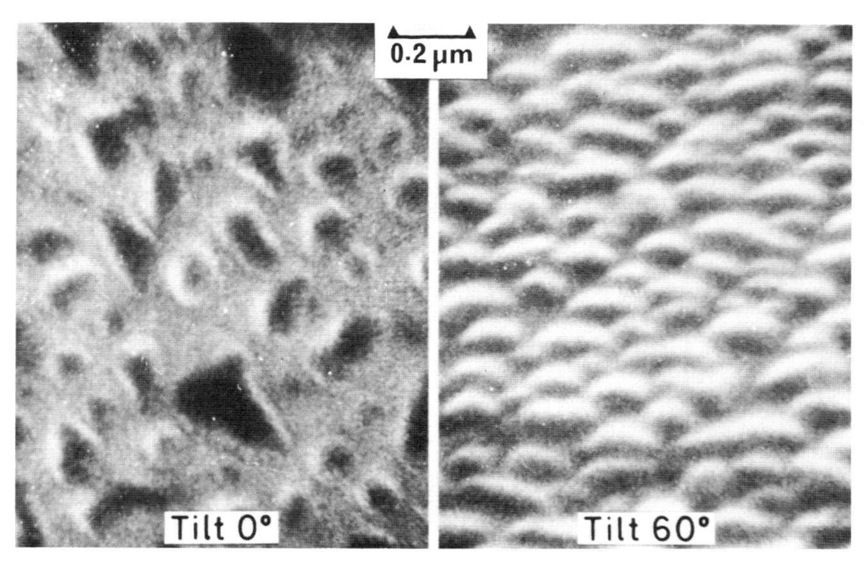


Fig. 10. Si growth centers formed at a deposition rate of 1 Å $s^{-1}$ with a diamond substrate temperature of 900 °C. (Scanning electron micrographs, secondary emission mode.)

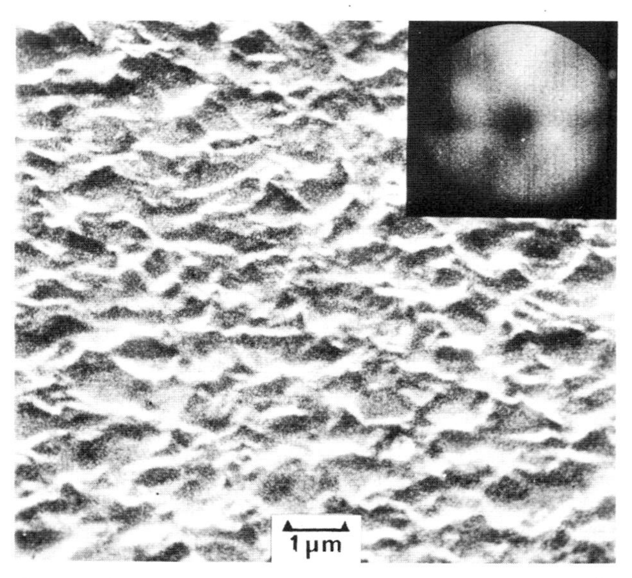


Fig. 11. Si film approximately 0.4 µm in thickness grown at 1 Å $s^{-1}$ with a diamond substrate temperature of 900 °C. In the inset the selected area electron channelling pattern exhibits a weak [111] pole. (Scanning electron micrograph, secondary emission mode.)

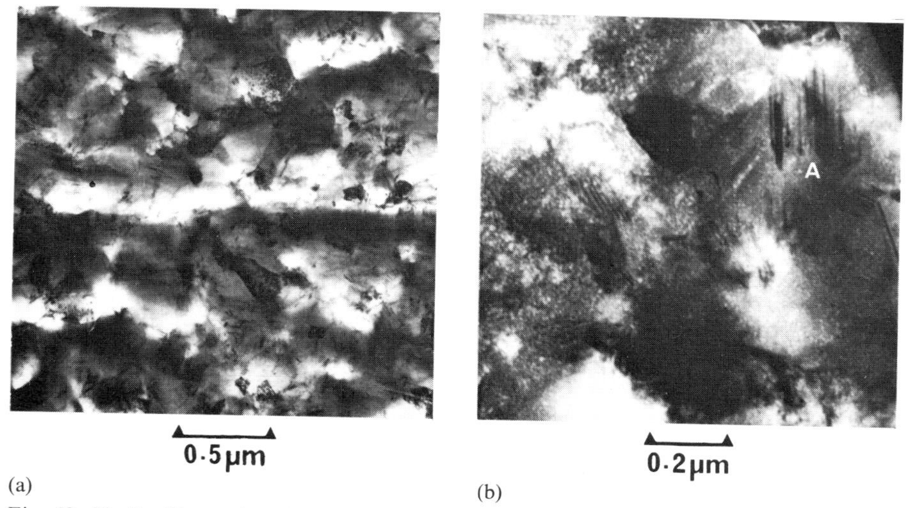


(a) (b)

Fig. 12. Similar film to that shown in Fig. 11: (a) a region near the surface illustrating the uneven structure; (b) a thicker region illustrating the highly defective nature and, for example, microtwins near A. (Transmission electron micrographs, 1 MeV TEM, bright field.)

## 4. DISCUSSION

### *4.1. Silicon on sapphire*

Sapphire is a most useful insulating substrate material for Si film growth since it is readily available as large single crystals with a high crystallographic perfection. However, it has a trigonal lattice, so that the mismatch with single-crystal Si depends upon their relative orientations and is generally quite large. Also, the materials differ in their coefficients of thermal expansion by a factor of approximately two[1].

The epitaxial relationship observed during the present work (*i.e.* $(111)_{Si}\|(0001)_{Al_2O_3}$ and $[1\bar{1}0]_{Si}\|[11\bar{2}0]_{Al_2O_3}$) was originally determined by Manasevit *et al.*[5] This may be rationalized, for example by use of the model constructed by Larssen[23], if substrate surface reconstruction[13] at the growth temperature is neglected. Furthermore, the difference between the magnitudes of the Si $\frac{1}{2}[1\bar{1}0]$ and sapphire $\frac{1}{3}[11\bar{2}0]$ lattice vectors is about 20% (room temperature value), so that closely spaced (~20 Å) misfit dislocations might be expected to be present at the interface. Arrays of such dislocations have been observed, for example, at metal[24] and semiconductor[25] heterojunctions. Nevertheless, the experimental observations presented above did not provide definite evidence for the presence of large-scale dislocation networks in the present system. Under certain TEM diffraction conditions, periodic contrasts with spacings of about 30–70 Å were observed. However, these were generally only weakly visible and the most likely origin of those with the smaller spacings would seem to be a moiré effect. Nevertheless, the occurrence of the larger spacing fringes and the occasional observation of more prominent and less regular contrasts in certain localized regions suggest that some small dislocation arrays may have been present, although the basic moiré fringe/dislocation contrast ambiguity[26] renders interpretation somewhat uncertain. Use of the weak-beam method of microscopy[27] might resolve such a problem, although in the present experiments it was found to be difficult to establish the required electron diffraction conditions due, in part, to film distortions produced by residual strains.

It is important to observe that mottled image contrast features were invariably given both by Si–sapphire interface regions and by exposed sapphire substrate areas which had accumulated no Si growth. These irregular features may be taken to indicate that a reaction probably occurred between the Si and the sapphire during the deposition process, so that the substrate surface became roughened. Other workers[7,13] have proposed certain reaction schemes involving the formation of volatile oxides (SiO and $Al_2O$) and Al. However, it also seems likely that a solid residue of an aluminosilicate (such as sillimanite, $Al_2SiO_5$) might be formed by a reaction of the type

$$Si+2Al_2O_3 \rightarrow Al_2SiO_5+Al_2O\uparrow$$

Therefore, it is not unreasonable to suppose that a small amount of this compound oxide could have been present at the Si–sapphire interfaces and possibly could have interfered with the formation of misfit dislocation arrays. It is also interesting to note that edge dislocations formed at the Si–sapphire interface during film growth could have been preferential sites for a solid state reaction, due to the presence of atoms with unsatisfied bonding potential at their cores. In this way, interfacial dislocations might be structurally unstable under growth conditions. However, general chemical interaction between Si and sapphire should be minimized by film growth at high deposition rates, and other workers are currently investigating interfaces produced in this manner[28].

Initial Si film growth was inferred to have occurred by nucleation since the thinnest films examined were composed of separate three-dimensional growth centers, in agreement with, for example, previous surface replica studies[12]. The number density of the larger growth centers formed at 900 °C was in the region of $10^9$ $cm^{-2}$ and, since considerable coalescence had already occurred, the initial saturation number density was considerably greater. It is interesting to note that such a value is more closely in accord with the predictions of conventional nucleation theory than the observed number densities of Si centers produced by growth upon Si substrates[19,29], where substantially lower numerical values have been found. Linington[30] has studied growth center coalescence using computer modelling techniques.

The thinnest Si films were composed of two 60° rotational twins, each with a relationship of the type $[1\bar{1}0]_{Si} \| [11\bar{2}0]_{Al_2O_3}$. Since the smallest growth centers were composed of only one of the twins, it is clear that initial "double positioning"[31] had taken place. Previous investigators[12,32] have also reported the formation of rotational twins having relationships of the type $[1\bar{1}0]_{Si} \| [1\bar{1}00]_{Al_2O_3}$, although no such twinning was observed in the present experiments.

Thin Si films grown at the higher temperature (1100 °C) were observed to contain especially large numbers of planar defects (stacking faults and microtwin lamellae). Since high temperatures would favor chemical interaction between the Si and the sapphire leading to enhanced formation of products such as $Al_2SiO_5$, it is possible that many of the planar defects were produced when the Si film grew over residues of this type. The mechanism of defect formation could be similar to that proposed by Booker and Stickler[33] for stacking fault generation in epitaxial Si films on Si substrates. It is also possible, of course, that some of the film defects were introduced at the boundaries of initial islands by imperfect coalescence.

In general, as the Si film thickness increased from a few hundred ångströms, TEM examination showed that further microtwins were introduced together with additional misorientations. Although the thickest film examined (2.9 μm) had a quite uniform profile, it still contained many defects including relatively large blocks of misoriented Si in its outer regions. Several factors may have been important in defect production. As final channels in the original deposit were grown over, general disregistries and accumulated Si–sapphire reaction products may have introduced imperfections. Also, during initial cooling after deposition differential Si/substrate thermal contraction would have introduced stresses which might have been partially relieved by defect generation. Indeed, many of the thin sapphire substrates were observed to deform physically during cooling. While details of defect depth distributions were not studied in the present investigation, other workers[34] have recently employed a film cross-sectioning technique in TEM investigations. They directly observed an improvement in film perfection with increasing distance from the Si–sapphire interface for films grown by $SiH_4$ pyrolysis.

*4.2. Silicon on diamond*

While type IIa diamond crystals, for example, are electrically insulating, their thermal conductivity[35] is approximately five times that of Cu. This combination of physical properties is potentially advantageous for devices fabricated in Si grown directly on such material. However, although Si and diamond have the same crystal structure, the diamond lattice parameter is 3.567 Å, so that the Si/diamond misfit at room temperature is about 34.4%. Also, the thermal expansion coefficients[36] of the two materials are somewhat different.

Due to the great resistance of diamond to chemical attack, in the present experiments it was considered that the optimum (111) substrate surface preparation was by cleavage. Although such a surface was expected to be covered with steps, it would have been unwise to remove these by mechanical abrasion since this would have produced severe surface damage on a fine scale[37] which could not then have been removed by a convenient chemical method. It is also important to note that diamond transforms to graphite in vacuum at temperatures at least as low as 1500 °C[38]. Since this transformation is initiated at the surface, it could have extremely deleterious effects on subsequent epitaxial growth. Therefore high temperature substrate "cleaning" pretreatment would seem to be of dubious usefulness. It was hoped that the transformation could be neglected at the substrate temperature chosen for Si deposition (900 °C).

Initial film growth was by nucleation giving a high number density of three-dimensional growth centers. The triangular shapes sometimes exhibited suggest a definite crystallographic relationship with the substrate. The centers overlapped to give a fairly uniform film at 0.4 μm thickness which at least partially preserved the (111) orientation of the substrate, although it contained a very large number of defects. It is clear that the defects will degrade the electrical properties of the Si and it is important to determine the structures of thicker films which might be used for device fabrication. It is also important to note, however, that autodoping effects similar to those found for sapphire substrates[7] will not occur.

## 5. CONCLUSION

Si deposition onto (0001) sapphire substrates gave initial growth by nucleation and three-dimensional growth center formation. The film orientation was that expected[5] ($(111)_{Si} \| (0001)_{Al_2O_3}$) and initial growth centers were directly observed to be composed of one or other of two 60° rotation twins, each with $[1\bar{1}0]_{Si} \| [11\bar{2}0]_{Al_2O_3}$.

While no definite evidence was obtained for large-scale arrays of misfit dislocations at the Si–sapphire interface, fringe image contrast was observed in the TEM. The weaker fringes of smallest spacing were thought to have been of moiré origin, while others including less regular and stronger contrasts in certain localized areas may have been dislocation images. However, there was evidence for a general substrate surface roughening which was thought to indicate that a reaction had taken place, possibly leaving residues of the type $Al_2SiO_5$. Such a reaction product could have impeded the formation of misfit dislocation arrays and also have initiated the formation of the planar defects, such as stacking faults, which were often found in the Si films, especially those grown at the higher temperature (1100 °C). When film thickness was increased, planar defects were still found to be present, together with other misoriented growth structures. The thickest film examined (2.9 μm) contained relatively large blocks of misoriented Si. These general observations were rationalized on the basis of impurity effects, disregistries between growth centers and strains introduced by thermal contraction during initial cooling.

The growth of Si on (111) diamond proceeded by initial nucleation leading, once again, to three-dimensional growth center formation. The centers overlapped to give a relatively uniform film which exhibited significant retention of the substrate orientation. Such films contained many defects including faults and microtwins.

## ACKNOWLEDGMENTS

This paper is based upon work submitted in partial fulfilment of the requirements for the degree of D. Phil. at the University of Oxford, 1972.

The authors would like to thank Dr. E. A. D. White (Imperial College) for supplying the sapphire substrates, Professor P. B. Hirsch for providing laboratory facilities and the Plessey Company Ltd., for giving financial support.

REFERENCES


1 J. D. Filby and S. Nielsen, *Br. J. Appl. Phys.*, *18* (1967) 1357.
2 G. W. Cullen, *J. Cryst. Growth*, *9* (1971) 107.
3 B. A. Joyce, *Rep. Prog. Phys.*, *37* (1974) 363.
4 H. M. Manasevit, *J. Cryst. Growth*, *22* (1974) 125.
5 H. M. Manasevit, A. Miller, F. L. Morritz and R. L. Nolder, *Trans. Metall. Soc. AIME*, *233* (1965) 540.
6 R. L. Nolder, D. J. Klein and D. H. Forbes, *J. Appl. Phys.*, *36* (1965) 3444.
7 R. W. Bicknell, B. A. Joyce, J. H. Neave and G. V. Smith, *Philos. Mag.*, *14* (1966) 31.
8 M. Tamura and M. Nomura, *Appl. Phys. Lett.*, *11* (1967) 196.
9 C. A. T. Salama, T. W. Tucker and L. Young, *Solid-State Electron.*, *10* (1967) 339.
10 C. T. Naber and J. E. O'Neal, *Trans. Metall. Soc. AIME*, *242* (1968) 470.
11 L. R. Weisberg and E. A. Miller, *Trans. Metall. Soc. AIME*, *242* (1968) 479.
12 Y. Yasuda and Y. Ohmura, *Jpn. J. Appl. Phys.*, *8* (1969) 1098.
13 C. C. Chang, *J. Vac. Sci. Technol.*, *8* (1971) 500.
14 D. J. Dumin and P. H. Robinson, *J. Appl. Phys.*, *39* (1968) 2759.
15 S. T. Picraux, *Appl. Phys. Lett.*, *20* (1972) 91.
16 C. B. Norris, *Appl. Phys. Lett.*, *20* (1972) 187.
17 P. F. Linington, *Proc. 25th Anniv. Meeting E.M.A.G., Cambridge, 1971*, Inst. Phys., London, p. 182.
18 H. M. Manasevit, F. M. Erdmann and A. C. Thorsen, *Proc. 146th E.C.S. Meeting, New York, 1974*, Electrochem. Soc., Princeton, N.J., p. 332.
19 A. G. Cullis and G. R. Booker, *J. Cryst. Growth*, *9* (1971) 132.
20 G. R. Booker, *Discuss. Faraday Soc.*, *38* (1964) 298.
21 G. R. Booker, in S. Amelinckx, R. Gevers, G. Remaut and J. van Landuyt (eds.), *Modern Diffraction and Imaging Techniques in Material Science*, North-Holland Publ. Co., Amsterdam, 1970, p. 613.
22 S. M. Davidson, G. R. Booker and R. Stickler, *Proc. 25th Anniv. Meeting E.M.A.G., Cambridge, 1971*, Inst. Phys., London, p. 298.
23 P. A. Larssen, *Acta Crystallogr.*, *20* (1966) 599.
24 J. W. Matthews, *Proc. Conf. on Single Crystal Films, Pennsylvania, 1963*, Macmillan, New York, p. 165.
25 A. G. Cullis and G. R. Booker, *Proc. 25th Anniv. Meeting E.M.A.G., Cambridge, 1971*, Inst. Phys., London, p. 320.
26 A. R. Thölén, *Phys. Status Solidi A*, *2* (1970) 537.
27 D. J. H. Cockayne, I. L. F. Ray and M. J. Whelan, *Philos. Mag.*, *20* (1969) 1265.
28 M. S. Abrahams, personal communication, 1975.
29 B. A. Joyce, R. R. Bradley and G. R. Booker, *Philos. Mag.*, *15* (1967) 1167.
30 P. F. Linington, *Ph.D. Thesis*, University of Cambridge, England, 1975.
31 E. W. Dickson and D. W. Pashley, *Philos. Mag.*, *7* (1962) 1315.
32 T. L. Chu, M. H. Francombe, G. A. Gruber, J. J. Oberly and R. L. Tallman, *Westinghouse Res. Lab. Rep. No. AFCRL-65-574, AD61992*, 1965.
33 G. R. Booker and R. Stickler, *J. Appl. Phys.*, *33* (1962) 3281.
34 M. S. Abrahams and C. J. Buiocchi, *Appl. Phys. Lett.*, *27* (1975) 325.
35 R. W. Ditchburn and J. F. H. Custers, in R. Berman (ed.), *Physical Properties of Diamond*, Clarendon Press, Oxford, 1965, p. 295.
36 J. Thewlis and A. R. Davey, *Philos. Mag.*, *1* (1956) 409.
37 D. Driver, E. M. Wilks and J. Wilks, *Proc. Int. Conf. on Synthetic Diamonds, Kiev, 1971*, p. Cl.
38 T. Evans and P. F. James, *Proc. R. Soc. London, Ser. A*, *277* (1964) 260.

*Thin Solid Films*, *31* (1976) 69–88



# ELECTRICAL CHARACTERIZATION OF EPITAXIAL LAYERS

G. E. STILLMAN* AND C. M. WOLFE**
*Lincoln Laboratory, Massachusetts Institute of Technology, Lexington, Mass. 02173 (U.S.A.)*
(Received April 3, 1975; accepted June 17, 1975)

The techniques for determining the concentrations of donors and acceptors in semiconductor samples from Hall effect and resistivity measurements are described, using measurements on GaAs as an example. Analyses of the temperature variation of the carrier concentration and mobility permit the determination of $N_D$ and $N_A$ in the range $1\times 10^{12} \lesssim N_D+N_A \lesssim 3\times 10^{17}$ cm$^{-3}$. An empirical curve is derived from the results of these analyses which permits the accurate determination of $N_D$ and $N_A$ in homogeneous GaAs samples from a single measurement of the Hall coefficient and resistivity at 77 K.

---

## INTRODUCTION

Electrical characterization can give considerable information about the purity of epitaxial layers. Such information is important for the evaluation and control of crystal growth procedures used to prepare high quality epitaxial material for device applications. It is also important for the detailed characterization of epitaxial material in conjunction with other evaluation procedures or in the interpretation and correlation of data from various experiments. Included under the term electrical characterization procedures are Hall coefficient and resistivity measurements, capacitance–voltage measurements, photoconductivity, photo-Hall and photo-capacitance measurements and other thermoelectric and galvanomagnetic transport measurements. All of these measurements can be useful for the characterization of particular materials. However, analysis of Hall coefficient and resistivity measurements provides more information about the purity of epitaxial layers than any of the other characterization methods. Carefully controlled doping experiments and photoconductivity and/or photoluminescence measurements can permit the identification of the particular impurities present in a sample, and the concentration of these impurities can best be determined from the analysis of carrier concentration *versus* temperature or mobility *versus* temperature data.

---

* Present address: Department of Electrical Engineering and Materials Research Laboratory, University of Illinois, Urbana, Ill. 61801, U.S.A.
** Present address: Department of Electrical Engineering and Laboratory for Applied Electronic Sciences, Washington University, St. Louis, Mo. 63130, U.S.A.

In this paper, using epitaxial GaAs as an example, we first discuss the analysis of Hall coefficient data as a function of temperature to determine the donor and acceptor concentrations in the epitaxial layers. Next, a method for evaluating the donor and acceptor concentrations from mobility data in samples in which the temperature variation of the Hall coefficient is not large enough for the usual analysis to be performed is described, and the effect of various scattering mechanisms on the calculated Hall mobility is examined. Finally, using the results of Hall coefficient and mobility analyses, a routine method for estimating the donor and acceptor concentrations from a single measurement of the Hall coefficient and resistivity at liquid nitrogen temperature is described.

Although the specific measurements and analyses referred to here are for high purity GaAs, the same considerations will be important for the measurement and analysis of Hall coefficient and resistivity on any high purity epitaxial layer.

## EXPERIMENTAL TECHNIQUES

The experimental techniques used for the measurement of the Hall coefficient and resistivity of semiconductor samples are well known. For measurements of epitaxial layers, the substrate must be either of sufficiently high resistivity not to cause a shunting effect on the epitaxial layer, or thick enough for the substrate to be removed before the measurements are made. For applications where high purity n-type layers are required on $p^+$ substrates, the measurements can be made using the depletion width of the p–n junction for isolation, provided that the currents and voltages used for the measurements are chosen so that the change in the depletion width does not influence the measurements. Where high purity n-type layers on $n^+$ substrates must be characterized, it has been found that with GaAs the layer can be accurately characterized by simultaneously growing an epitaxial layer on an adjacent high resistivity substrate and then using this sample for the characterization measurements. High resistivity GaAs substrates ($\rho \gtrsim 10^7\ \Omega$ cm at room temperature) are available which permit measurements on the highest purity GaAs obtainable without the substrate having any influence on the electrical measurements.

To obtain the most information from the Hall coefficient and resistivity measurements, they should be made over a wide temperature range—for GaAs, over the temperature range 4.2 K $\lesssim T \lesssim$ 400 K. Interpretation of measurements on GaAs at higher temperatures must include the influence of the higher conduction band minima, and such measurements contribute little additional information about the purity of the sample. Measurements at temperatures less than 4.2 K also contribute very little additional information about the purity of the epitaxial layer. For measurements in the lower part of the temperature range especially, precautions must be taken to ensure, as far as possible, that the sample is at a uniform temperature, that the background is at the same temperature as the sample, and that the actual temperature of the sample is accurately measured. The measurement apparatus must be designed so that the capacitance is low enough to achieve reasonable time constants with the high sample resistances which are encountered at low temperatures. The leakage resistance of the sample holder, leads, switches etc. must also be negligible compared with the sample resistance.

The measurement method should be chosen to eliminate the unwanted voltages which appear at the Hall terminals due to *IR* drops and the various thermoelectric and galvanomagnetic effects. By taking four d.c. measurements of the Hall voltage at each temperature with two directions of the magnetic field and sample current, it is possible to eliminate all of the error voltages except the Ettingshausen voltage (which can be eliminated by a.c. techniques if necessary) provided that the time constants are short enough for the measurements to be completed before the thermal gradients due to the Peltier and Righi–Leduc effects have had time to change or reverse[1]. In practice this becomes difficult to do at low temperatures because the thermal time constants generally decrease with decreasing temperature while the electrical time constant (sample resistance) increases.

Another complication which must be considered in the measurements at low temperatures is the non-ohmic behavior observed in many semiconductors. In n-type GaAs, this non-ohmic behavior results from impact ionization of the shallow impurity levels[2,3], and the Hall coefficient and resistivity measurements must be made at currents which are low enough for the Hall and resistivity voltages to be linear with the corresponding currents.

When the precautions just discussed are carefully observed, it is possible to measure the Hall and resistivity voltages accurately over a wide temperature range using either standard or van der Pauw geometries, and with the appropriate sample dimensions the Hall coefficient and resistivity can be accurately determined from these measurements.

## DETERMINATION OF CARRIER CONCENTRATION AND MOBILITY

The parameters determined from the experimental measurements are the Hall coefficient and resistivity, while the quantities of interest for the determination of the electrically active impurity concentrations are the free carrier concentration and the free carrier mobility. The carrier concentration $n$ is determined from the measured Hall coefficient $R_H$ using the expression

$$n = r_H/eR_H \tag{1}$$

in which $e$ is the electronic charge and $r_H$, the Hall coefficient factor, is a numerical factor close to unity which relates the carrier concentration to the measured Hall coefficient.

The Hall coefficient factor can be calculated using the distribution function obtained from solution of the Boltzmann equation for the applied electric and magnetic fields, with the appropriate carrier scattering. Hall coefficient factors for various scattering mechanisms have been discussed in detail by Beer[4]. For scattering processes in which the change in energy is small, a relaxation time approximation can be used for the solution of the Boltzmann equation. The calculated Hall coefficient factors in the relaxation time approximation are $r_H = 3\pi/8 = 1.18$ for acoustic mode deformation potential scattering, $r_H = 45\pi/$

$128 = 1.10$ for acoustic mode piezoelectric scattering, $1 \lesssim r_H \lesssim 315\pi/512 = 1.93$ for ionized impurity scattering and $r_H = 1$ for neutral impurity scattering[4]. When the scattering process is such that the energy change of the scattered carriers is relatively large compared with their initial energy, a universal relaxation time cannot be defined, and the Boltzmann equation must be solved by other techniques. Scattering by optical phonons through the polar interaction is one such process which is important for GaAs and other III–V compound semiconductors. The Hall coefficient factor for polar mode scattering has been measured experimentally[5] and studied theoretically by several authors[6–8].

The measured Hall coefficient and resistivity can be combined to calculate the Hall mobility $\mu_H = R_H/\rho$. The conductivity mobility is then given by $\mu = \mu_H/r_H$. The Hall coefficient factor $r_H$ is in general a function of the magnetic field used for the measurement as well as of the sample degeneracy and temperature. Experimentally, the Hall coefficient factor at a given magnetic field $B$ can be determined from $r_H(B) = R_H(B)/R_H(\infty)$, in which $R_H(\infty)$ is the Hall constant in the high magnetic field limit. By making the measurements in the high magnetic field limit, where the Hall coefficient factor is unity, the free carrier concentration can be determined directly from the experimental measurement of the Hall coefficient, and, if the temperature is not so low and the magnetic field not so high as to influence significantly the energy level structure of the impurity levels and/or the conduction and valence bands (and therefore the freeze-out of the free carriers), it can then be used to determine $n$. Experimentally, it is not usually possible to achieve these conditions routinely over a wide temperature range. Nevertheless, in most reported analyses of Hall coefficient and mobility data to determine the ionized impurity concentration of epitaxial layers, it is assumed that $r_H = 1$. The effect of this assumption must be considered in estimating the accuracy of the values of $N_D$ and $N_A$ determined from such an analysis of carrier concentration *versus* temperature data, particularly when the Hall coefficient factor is also a function of temperature.

## DETERMINATION OF DONOR AND ACCEPTOR CONCENTRATIONS

### *Carrier concentration* versus *temperature analysis*

To determine the donor and acceptor concentrations from the temperature variation of the carrier concentration, a theoretical carrier concentration equation must be fitted to the experimental carrier concentration determined from eqn. (1). For the usual model of a non-degenerate n-type semiconductor with a shallow donor concentration $N_D$ and acceptor concentration $N_A$, the acceptors are fully ionized at all temperatures of interest and the concentration $n_0$ of free electrons in the conduction band is given by the solution of

$$\frac{n_0(n_0+N_A)-n_i^2}{N_D-N_A-n_0-(n_i^2/n_0)} = \frac{N_c}{g_1}\exp\left(-\frac{E_D}{kT}\right) \tag{2}$$

where $N_c = 2(2\pi m_D^* kT/h^2)^{3/2}$. In this equation, $n_i$ is the intrinsic carrier concentration, $n_i^2 = n_0 p_0$, $m_D^*$ is the conduction band density-of-states effective mass, $g_1$

is the degeneracy of the ground state of the impurity center, and $E_D$ is the energy of the ground state below the conduction band; all excited states have been neglected. The equation can readily be extended to more complicated semiconductor models and arbitrary degeneracy conditions[9]. In particular temperature ranges, it is possible to approximate eqn. (2) by simpler relations, and these approximations have been utilized by some workers to estimate the donor and acceptor concentrations without resorting to a detailed fit of the theoretical equation to the experimental data:

(1) at very low temperatures, where $n_0 \ll N_A$, $N_D - N_A$ and $p_0 = n_i^2/n_0 \approx 0$,

$$n_0 \approx \frac{N_c}{g_1} \frac{(N_D - N_A)}{N_A} \exp\left(-\frac{E_D}{kT}\right) \tag{3}$$

(2) at slightly higher temperatures or for low values of $N_A$, where $n_0 \gg N_A$ and $p_0 \approx 0$,

$$n_0 \approx \left\{\frac{N_c}{g_1}(N_D - N_A)\right\}^{1/2} \exp\left(-\frac{E_D}{2kT}\right) \tag{4}$$

(3) at still higher temperatures (exhaustion region), where $E_D \ll kT$ but $n_0 \gg n_i$,

$$n_0 \approx N_D - N_A = \text{constant} \tag{5}$$

By plotting the experimental carrier concentration *versus* $1/T$ on semi-log graph paper, simple estimates can be made of $E_D$ and $(N_D - N_A)/N_A$ from the slope and intercept of the straight line fitted to the data in the appropriate temperature range using eqns. (3)–(5). Using the value of $n_0 = N_D - N_A$ in the exhaustion temperature range (eqn. (5)), the donor and acceptor concentrations can both be determined. This procedure has the advantage of being simple, but it generally neglects the temperature dependence of $N_c$. For GaAs with $N_D - N_A$ in the range $10^{14}$–$10^{16}$ cm$^{-3}$, this can introduce considerable error.

A more accurate procedure, and one that is nearly as simple, is to use a computer or programmable calculator to fit eqn. (2) to the experimental data by adjusting $N_D$, $N_A$ and $E_D$, or some other combination of these variables, to minimize the error between the calculated value $n_0$ and the experimentally determined value $n$ of the free carrier concentration. To weight the data points equally over the entire temperature range, the function which is mimimized is usually some root mean square, such as $[\Sigma_j\{\log(n_{0_j}/n_j)\}^2]^{1/2}$, in which the sum is over the various experimental points and $n_{0_j}$ is the calculated free carrier concentration at the temperature $T_j$ corresponding to the experimental value $n_j$ at the same temperature.

Equations (1) and (2) have been used successfully by several workers for the analysis of high purity GaAs Hall coefficient data. Bolger *et al.*[10,11] used these equations, neglecting the terms involving $n_i^2$ and assuming $r_H = 1$, to analyze Hall coefficient data taken at a magnetic field of 5 kG with good results. The value used for $m_D^*$ was 0.072 m, and the value used for the degeneracy factor $g_1$ was 2.

Stillman *et al.*[12] also used these same equations and values for $r_H$, $m_D^*$ and $g_1$ for the analysis of Hall coefficient data taken at a magnetic field of 5 kG. The experimental points and the calculated curves, using the values of $N_D$, $N_A$ and $E_D$ in Table I, for six different samples are shown in Fig. 1. The values of $N_D$, $N_A$ and $E_D$ were determined from a least-squares fit of eqn. (2) to the experimental data for each sample. The deviation of the experimental data from the calculated curves at low temperatures for samples 1–3 is due to two-band conduction involving the conduction band and an impurity band which results from overlap of the impurity ground state wavefunctions in these more heavily doped samples. The values of $E_D$, determined from the fit to the experimental data at temperatures higher than those where impurity band conduction becomes significant, decrease rapidly with increasing donor concentration from greater than 5.5 meV, for the highest purity samples, to less than 1.9 meV for $N_D = 4.7 \times 10^{15}$ cm$^{-3}$.

Equations (1) and (2) and the same values of $r_H$, $m_D^*$ and $g_1$ were also used by Maruyama *et al.*[13] and again good fits to the experimental data were obtained.

In these three cases[10–13], the fit to the experimental data was sufficiently good to justify the neglect of excited states in the calculation in view of the approximation $r_H = 1$. Eddolls *et al.*[14, 15], however, found that these equations and parameters could only give a good fit to the experimental data on high purity vapor-phase epitaxial GaAs for temperatures below 20 K. He observed a minimum in the $R_H$ *versus* $1/T$ curve at a temperature of about 80 K in Hall coefficient measure-

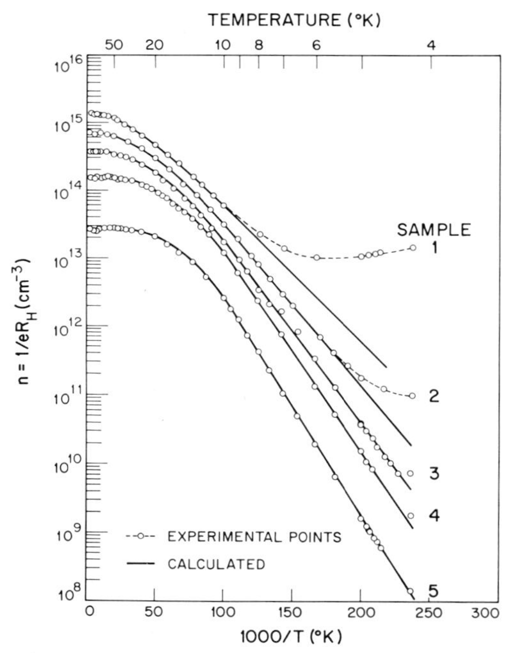


Fig. 1. Experimental and calculated variation in the electron concentration with temperature for six GaAs samples. The curves were calculated using eqn. (2) and the parameters given in the text and in Table I.

TABLE I

PROPERTIES OF GaAs SAMPLES

| *Sample* | $N_D$ $(cm^{-3})$ | $N_A$ $(cm^{-3})$ | $\mu_{77\,K}$ $(cm^2\ V^{-1}\ sec^{-1})$ | $E_D$ $(10^{-3}\ eV)$ |
|---|---|---|---|---|
| 1 | $4.72\times10^{15}$ | $1.61\times10^{15}$ | 33 800 | 1.89 |
| 2 | $2.06\times10^{15}$ | $6.78\times10^{14}$ | 46 300 | 3.29 |
| 3 | $1.06\times10^{15}$ | $3.27\times10^{14}$ | 72 000 | 3.88 |
| 4 | $5.02\times10^{14}$ | $1.36\times10^{14}$ | 107 000 | 4.51 |
| 5 | $2.04\times10^{14}$ | $4.07\times10^{13}$ | 153 000 | 5.09 |
| 6 | $4.80\times10^{13}$ | $2.13\times10^{13}$ | 210 000 | 5.52 |

ments which were made with a magnetic field of 3 kG, similar to the behavior observed by Stillman *et al.*[5] Eddolls *et al.* attributed the minimum to the magnetic field dependence of $R_H$, but in this case the minimum disappeared when a small magnetic field was used (465 G), in contrast to the variation expected for polar mode scattering and that observed experimentally by Stillman *et al.*[5] By including excited states in the calculation, it was possible to obtain a good fit to the experimental data over the entire temperature range. Carballés *et al.*[16] also found it necessary to include excited states in order to fit their experimental Hall coefficient data for high purity liquid-phase epitaxial GaAs.

The inclusion of excited states in the donor statistics was first considered by Shifrin[17] and Landsberg[18]. When the excited states are included, eqn. (2) is modified to (neglecting the terms in $n_i^2$)

$$\frac{n_0\,(n_0+N_A)}{N_D-N_A-n_0}=\frac{N_c\exp\,(-E_D/kT)}{g_1\,(1+F)} \tag{6}$$

where

$$F=\sum_{r=2}\frac{g_r}{g_1}\exp\left(-\frac{E_r}{kT}\right) \tag{7}$$

In these equations, $g_r$ is the degeneracy factor for the $r$th state of the impurity ($r = 1$ is the ground state, $r = 2$ is the first excited state etc.) and $E_r$ is the energy of the $r$th state measured above the energy of the ground state. The sum over $r$ in eqn. (7) is over all discrete excited states (*i.e.* omitting those excited states which are banded) of the impurity center.

Eddolls *et al.*[14, 15] and Carballés *et al.*[16] used eqns. (6) and (7) to include the effects of the excited states by assuming that the excited states were related to $E_D$ by the usual Bohr equation, so that $E_r = E_D\,(1-r^{-2})$. With the degeneracy factors from the hydrogenic model, the factor $1+F$ in eqn. (6) is given by

$$1+F=\sum_{r=1} r^2\exp\left\{-\frac{E_D\,(1-r^{-2})}{kT}\right\} \tag{8}$$

Although the fits to the experimental data obtained by Eddolls *et al.*[14, 15] and Carballés *et al.*[16] were much improved when the excited states were included using eqn. (8), it is now known that this is not the correct model for the energy level structure of the shallow donor levels, even though the shallow donor levels in GaAs are accurately hydrogenic[19].

The problem with the model for the excited states described above is the assumption that the relationship of the energy of the ground and excited states remains the same—*i.e.* the energy of the first excited state is always $\frac{3}{4}E_D$ above the ground state energy $E_D$ etc.—even though $E_D$ decreases rapidly with increasing impurity concentration. Far infrared photoconductivity measurements have in fact shown that the energy separation of the ground state and first excited state remains constant even when the thermal activation or ionization energy $E_D$ determined from the Hall coefficient analyses is smaller than the energy of the photoconductivity peak associated with this transition[12].

The far infrared extrinsic photoconductivity spectra of four GaAs samples with donor concentrations in the range $5 \times 10^{13}$–$2 \times 10^{15}$ cm$^{-3}$ are shown in Fig. 2. These spectra are shown on a relative scale only, and each curve has been shifted vertically to separate it from the other curves for clarity. The photoresponsivity of the samples decreases monotonically with increasing donor concentration. The dominant peak at 35.5 cm$^{-1}$ in each spectral curve has been identified with photoconductivity which results from excitation of electrons from the ground state of neutral shallow donors to the first excited state, followed by subsequent thermal excitation of these electrons into the conduction band continuum[20].

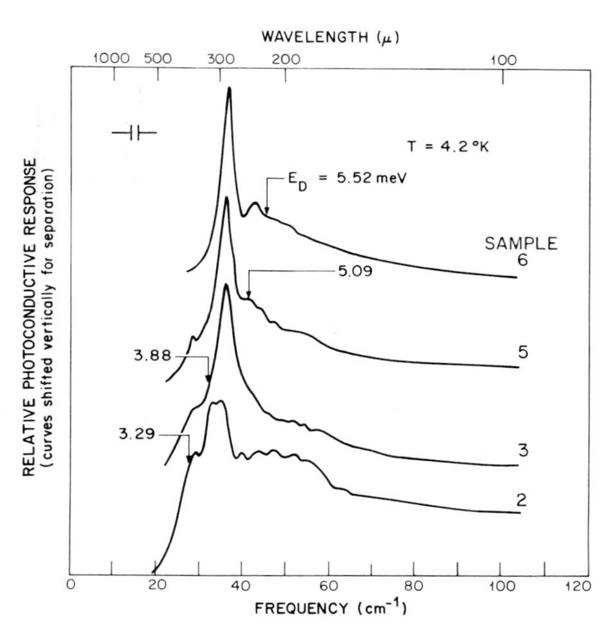


Fig. 2. Variation in GaAs far infrared extrinsic photoconductivity spectra with donor concentration. The samples are the same as those used in Table I and Fig. 1.

As the donor concentration increases, $E_D$ decreases, but the separation of the ground and excited states remains constant at 35.5 $cm^{-1}$ or 4.40 meV. This corresponds to a donor binding energy of about 5.86 meV. (More detailed measurements have shown that the dominant peak in each of the low resolution curves ($\sim$ $cm^{-1}$) of Fig. 2 actually consists of transitions from as many as four different impurity levels which have slightly different ground state energies due to central cell effects. Analysis of these data has given a hydrogenic binding energy for donors in GaAs of 5.715 meV [19].) For $N_D \approx 1 \times 10^{15}$ $cm^{-3}$ the thermal ionization energy is actually less than the energy separation of the ground and excited states. Other results show that the peak corresponding to the ground state–first excited state transition is still at 35.5 $cm^{-1}$ in the sample with $N_D = 2.06 \times 10^{15}$ $cm^{-3}$ and the apparent shift to lower energies is actually due to other transitions which presumably result from impurity interactions[21].

The constant separation of the ground and excited states while the thermal ionization energy decreases is consistent with a mechanism first proposed by Shifrin[22] to explain the decrease in the thermal ionization energy with increasing donor concentration. He reasoned that the decrease in the ionization energy with increasing impurity concentration was due to conduction in the overlapping excited states of the ionized donor atoms. Thus, the energy level structure of the donor levels consists of two parts, one having the discrete energy levels and the other a quasi-continuum of levels covering a range of energies $\Delta E$ below the conduction band. A diagram of this energy level structure for shallow donors in GaAs is shown in Fig. 3. The thermal ionization energy $E_D$ determined from Hall coefficient measurements is related to the isolated donor binding energy $E_I$ and the

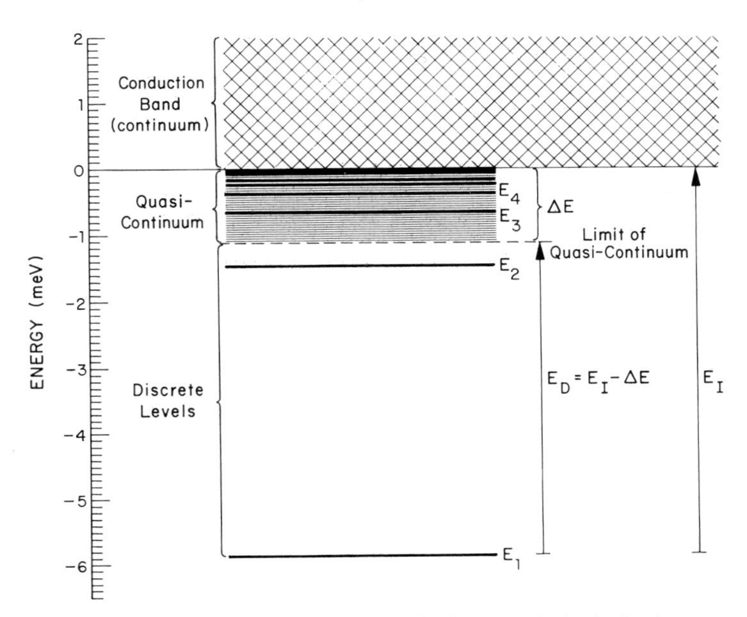


Fig. 3. Energy level structure for donors in GaAs. $E_I$ is the ionization energy of an isolated donor impurity.

width of the quasi-continuum $\Delta E$ by $E_D = E_I - \Delta E$. The variation in the width of the quasi-continuum determined using the thermal ionization energy obtained from the Hall coefficient analyses and the donor binding energy determined from the photoconductivity measurements is shown in Fig. 4. This figure shows that when the thermal ionization energy is less than about 4.3 meV and $N_D \gtrsim 8 \times 10^{14}$ cm$^{-3}$ no excited states should be included in the analysis. When the thermal ionization energy is less than about 5.1 meV but greater than 4.3 meV ($1 \times 10^{14} \lesssim N_D \lesssim 8 \times 10^{14}$), only one excited state should be included in the analysis. Thus, of the samples analyzed by Eddolls *et al.*[14,15], only one should have required the inclusion of excited states in the analysis, and then only one excited state should have been required. The reason that more excited states than this were necessary for the best fit to the experimental data is probably related to the temperature variation of the Hall coefficient factor in the high temperature range. The use of a model with none, one or two excited states can yield calculated total ionized impurity concentrations which differ by as much as 17%, while the corresponding ionization energies differ by less than 0.5%. Thus, for accurate estimates of the ionized impurity concentration, the model used, as well as the temperature variation of the Hall coefficient factor, must be carefully considered.

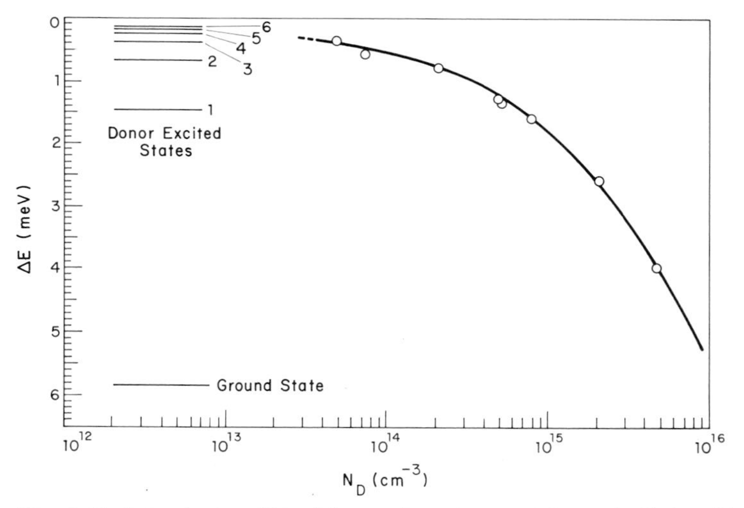


Fig. 4. Variation in the width of the quasi-continuum of donors in GaAs with donor concentration. The ground state and first six excited state energies of a shallow donor are shown for comparison.

*Mobility* versus *temperature analysis*

The analysis of the temperature variation of the Hall coefficient or carrier concentration is not a reliable method for determining the donor and acceptor concentrations in samples which do not have sufficient carrier freeze-out or which have significant impurity band conduction over most of the temperature range considered. For GaAs this occurs for donor concentrations higher than about $5 \times 10^{15}$ cm$^{-3}$. Some other method must be used to determine $N_D$ and $N_A$ in samples with higher values of $N_D$. In addition, there is considerable effort involved in taking Hall constant data over the required wide temperature range and then in analyzing the data to determine the donor and acceptor concentrations by the method described earlier. For routine evaluation of material or in other instances when it is not practical to use this procedure, it is desirable to be able to estimate the sample purity by some simpler method.

One method that has frequently been used to estimate $N_D$ and $N_A$ is to calculate these concentrations from the Brooks–Herring[23] mobility formula for ionized impurity scattering using the experimental mobility and carrier concentration measured at liquid nitrogen temperature (77 K). The Brooks–Herring equation for the ionized impurity scattering mobility $\mu_I$ is

$$\mu_I = \frac{3.28 \times 10^{15}\,(m/m^*)^{1/2}\,\varepsilon_0^2 T^{3/2}}{(2N_A + n)\,\{\ln(b+1) - b/(b+1)\}} \quad \mathrm{cm^2\,V^{-1}\,sec^{-1}} \tag{9}$$

where

$$b = \frac{1.29 \times 10^{14}\,(m^*/m)\,\varepsilon_0\,T^2}{n^*} \tag{10}$$

and $n^*$ is an effective screening density:

$$n^* = n + (n + N_A)\,(N_D - N_A - n)/N_D \quad \mathrm{cm^{-3}} \tag{11}$$

The rest of the terms have their usual meaning. For temperatures and/or samples in which donor de-ionization occurs, it is not possible to solve for $N_D$ and $N_A$ explicitly, and $N_D$ and $N_A$ must be adjusted separately to obtain agreement between the experimental and calculated mobilities. Where donor de-ionization does not occur, $n = N_D - N_A$ and so $n^* = n$. Thus the acceptor concentration $N_A$ can be calculated directly from the measured mobility and carrier concentration. In following this procedure the effects of lattice scattering on the mobility have usually been neglected and this leads to significant errors at low and intermediate impurity concentrations. The effect of lattice scattering can be included in an approximate way by assuming a lattice scattering mobility and simply combining it with the Brooks–Herring mobility. Although this procedure is somewhat better than neglecting the lattice scattering, it can still lead to significant errors in the intermediate impurity concentration range.

The Brooks–Herring equation has also been used to determine the impurity concentration from the experimental mobility measured at some lower temperature at which the effects of lattice scattering should be negligible[14, 15]. However, if the temperature is too low, other scattering mechanisms or impurity banding effects can have an effect on the measured mobility, whereas if the temperature is too high lattice scattering dominates. Therefore there is some optimum temperature at which the experimental mobility should be measured to make the determination of the impurity concentration most reliable. By performing this same type of analysis of the experimental mobility over the entire temperature range, effective values for the donor and acceptor concentrations $N_D^*$ and $N_A^*$ can be determined for each temperature and corresponding mobility. These values of the donor and acceptor concentrations are those that are required to give an ionized impurity mobility equal to the experimental mobility. If some other scattering mechanism or conductivity mechanism such as hopping conductivity or impurity band conduction is limiting the mobility, the calculated values of $N_D^*$ and $N_A^*$ will be larger than the actual values. The minimum values of $N_D^*$ and $N_A^*$ will occur in the temperature range in which the mobility is limited primarily by ionized impurity scattering and should place an upper bound on the actual values of $N_D$ and $N_A$. The temperature variation of the effective values $N_D^*$ and $N_A^*$ of the donor and acceptor concentrations determined in this way from experimental mobility measurements over the temperature range 300–4 K, for three different samples, is shown in Fig. 5[24]. Also shown in this figure are the values of $N_D$ and $N_A$ determined from analyses of Hall constant *versus* temperature measurements for these samples. The calculated values of $N_D^*$ and $N_A^*$ at high temperatures are much larger than the values of $N_D$ and $N_A$ determined from Hall constant analyses for all three samples because of the effects of lattice scattering mechanisms. The calculated values for the two less pure samples at very low temperatures are larger than the values from the Hall constant analyses because of impurity banding and/or hopping conduction effects which decrease

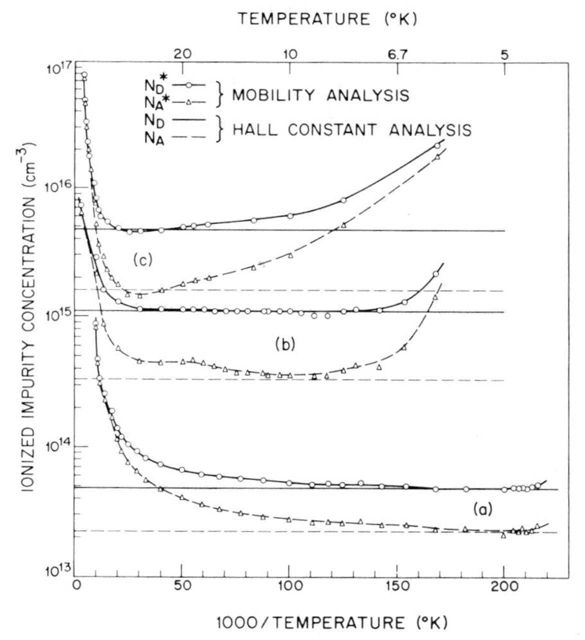


Fig. 5. Temperature variation in the effective donor and acceptor concentrations $N_D^*$ and $N_A^*$ for three different samples as determined from the analysis of the mobility *vs.* temperature data for these samples using the Brooks–Herring equation for ionized impurity scattering. The values of $N_D$ and $N_A$ determined from analysis of the carrier concentration *vs.* temperature data for each of these samples are shown by the straight lines. There is good agreement between the minimum values of $N_D^*$ and $N_A^*$ for each sample and the corresponding values of $N_D$ and $N_A$. The optimal temperature for the mobility analysis shifts from the regions marked (a) to (b) to (c) with increasing impurity concentration.

the experimental mobility. For the sample with the lowest impurity concentration, the values of $N_D^*$ and $N_A^*$ agree with the values of $N_D$ and $N_A$ even at the lower temperature, since for this sample the effects of hopping conduction or impurity banding are negligible even at the lower temperatures. There is very good agreement between the minimum values of $N_D^*$ and $N_A^*$ determined in this way and the values of $N_D$ and $N_A$ determined from the Hall constant analysis over the entire range of impurity concentration where analysis of Hall coefficient *versus* temperature data is reliable ($N_D+N_A \lesssim 7\times10^{15}$ cm$^{-3}$, $\mu_{77} \gtrsim 3\times10^4$ cm$^2$ V$^{-1}$ sec$^{-1}$).

The mobility analysis and the corresponding experimental measurements, when they are made over the entire temperature range, are not much easier to make than the standard Hall constant measurements and analysis, but they allow an optimum temperature to be selected for the mobility analysis, and more importantly this analysis can be extended to more heavily doped samples which do not show significant carrier freeze-out. The optimum temperature ranges for the mobility analysis for different ionized impurity concentrations and corresponding approximate 77 K mobility values are shown in Fig. 6. The vertical bars

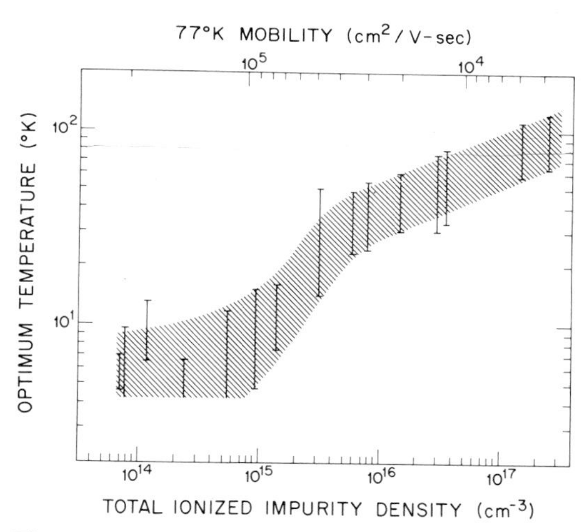


Fig. 6. Best temperature range for mobility analysis neglecting lattice scattering mechanisms. The vertical bars indicate the range of temperatures in which the total effective impurity density $N_D^*+N_A^*$ was within 10% of its minimum value.

in the figure show the temperature range in which the calculated value of $N_D^*+N_A^*$ is within +10% of its minimum value. Thus the shaded area in Fig. 6 represents the range of temperatures in which analysis of a single mobility measurement will yield a value for $N_D^*+N_A^*$ that is within 10% of the value that would be determined at the optimum temperature. These data show that, for samples with $\mu_{77\,K} \lesssim 12\,000$ cm$^2$ V$^{-1}$ sec$^{-1}$ up to the limit of the Born approximation, $N_D^*$ and $N_A^*$ can be determined from a single mobility measurement at 77 K. At this temperature $n^*$ is equal to the carrier concentration, so that eqns. (9)–(11) can be solved explicitly for $N_D+N_A$.

CALCULATED ELECTRON MOBILITY IN GaAs

By combining the techniques of the analysis of the temperature dependence of the carrier concentration and the analysis of the mobility at selected temperatures, the values of $N_D+N_A$ can be determined in the range from the lowest values attainable in the highest purity samples to values as high as about $3\times10^{17}$ $cm^{-3}$, above which there is no reason to expect the Brooks–Herring equation to be valid. A check on the accuracy of the values determined, as well as an evaluation of the importance of the various scattering mechanisms, can be made by calculating the carrier mobility including all the relevant scattering mechanisms.

For n-type GaAs the important electron scattering mechanisms are scattering by optical phonons through polar interactions[25], polar scattering due to piezoelectrically active acoustic phonons[26], deformation potential scattering by acoustic phonons[27], scattering by ionized impurities[23] and scattering by neutral impurities[28]. All of these scattering mechanisms, except the polar mode optical phonon scattering, can be accurately treated in the relaxation time approximation. The energy change of the electrons when scattered by the polar mode interaction with optical phonons can be large compared with their initial energy, so that a universal relaxation time cannot be defined for this scattering process. In general, some variational method must be used to solve the Boltzmann equation for arbitrary temperatures when this scattering mechanism is important. However, Ehrenreich[25] has shown that an effective relaxation time can be defined for this scattering mechanism over a restricted temperature range, which for GaAs is from 0 K to about 115 K. Using this approximation, Wolfe *et al.*[29] have calculated the electron mobility for high purity GaAs samples under the assumption that the effective relaxation time, when all the scattering processes are included, is given by

$$\frac{1}{\tau(x)}=\sum_i \frac{1}{\tau_i(x)} \tag{12}$$

where the $\tau_i(x)$ are the relaxation times for the individual scattering processes, the sum is over all the scattering processes, and $x$ is the electron energy in units of $kT$. The conduction band was assumed to be parabolic. Using classical statistics, the average relaxation time $<\tau>$ was calculated from

$$<\tau> = \frac{4}{3(\pi)^{1/2}}\int_0^\infty \tau(x)x^{3/2}\exp(-x)\,\mathrm{d}x \tag{13}$$

and the electron mobility was then determined from $\mu=e<\tau>/m^*$, where $e$ is the electronic charge and $m^*$ is the electron effective mass. The results of these calculations for sample 6 of Table I are shown in Fig. 7, along with the experimental mobility. Also shown in this figure are the calculated mobilities for each

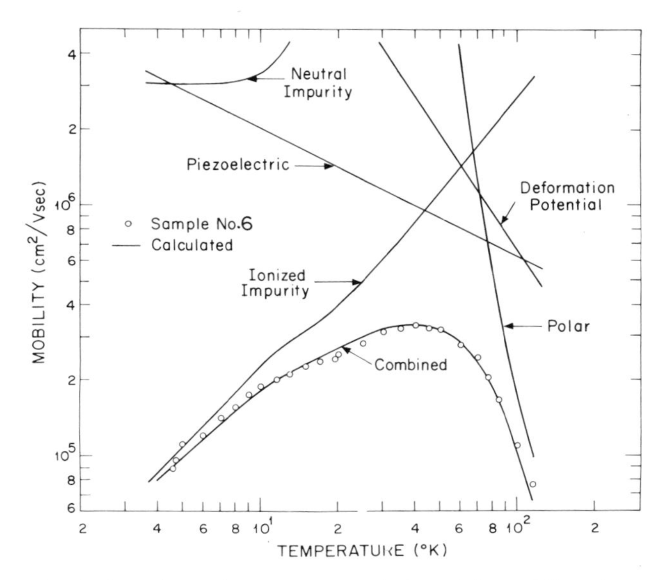


Fig. 7. Experimental temperature variation of the mobility of sample 6, Table I, and the calculated mobility curves for each scattering process acting separately and for all scattering processes combined.

of the scattering mechanisms acting separately. In the intermediate temperature range, all of the scattering processes are important. In particular, the calculated combined mobility in the temperature range 45–85 K varies by 5–6% when the deformation potential is varied by 10%. Thus the deformation potential was adjusted to obtain the best agreement between the experimental and calculated values of the mobility. The best agreement was obtained with a deformation potential of $|E_1| = 7.0$ eV. Figure 8 shows the calculated and experimental mobility variation for three different samples. The material parameters used for these calculated curves were the same as those used for the calculated mobility in Fig. 7, and the values of $N_D$, $N_A$ and $E_D$ determined for these samples from Hall constant analyses are given in the figure caption. The good agreement between the experimental and theoretical mobilities for these samples, particularly since no adjustable parameters were used, indicates that the values of $N_D$ and $N_A$ determined from Hall constant analysis are reasonably accurate.

Others workers have subsequently solved the Boltzmann equation for arbitrary magnetic field strengths, including polar mode optical phonon scattering and the elastic scattering mechanisms, without resorting to the relaxation time approximation. Fletcher and Butcher[30] have calculated the Hall mobility for the data of sample 6. These calculations were made for the same magnetic field as was used in the experimental measurements (5 kG). The agreement over the entire temperature range 4–400 K was extremely good. Other calculations have also given good agreement between the calculated mobility and experimental measurements on samples with slightly higher total ionized impurity concentrations[31,32].

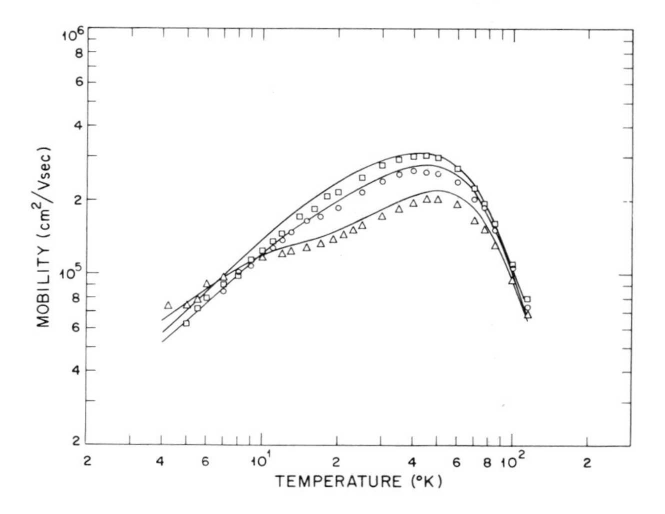


Fig. 8. Temperature dependence of the experimental mobility and calculated combined mobility curves for three samples with different values of $N_D$ and $N_A$: □, $N_D = 4.61 \times 10^{13}$, $N_A = 2.97 \times 10^{13}$, $E_D = 5.57$ meV; ○, $N_D = 7.15 \times 10^{13}$, $N_A = 3.66 \times 10^{13}$, $E_D = 5.31$ meV; △, $N_D = 1.91 \times 10^{14}$, $N_A = 2.61 \times 10^{13}$, $E_D = 5.30$ meV.

## IONIZED IMPURITY CONCENTRATION FROM HALL MEASUREMENT AT 77 K

The results of Hall coefficient and mobility analyses on a number of samples have been used to determine an empirical relationship between the electron Hall mobility measured at 77 K for a magnetic field of 5 kG and the total ionized impurity concentration[24]. The samples used had $N_D + N_A$ values between $7 \times 10^{13}$ and $3 \times 10^{17}$ cm$^{-3}$. Contrary to some misunderstanding in the literature[32], all the scattering mechanisms including lattice scattering are included in this empirical relationship for values of $N_D + N_A$ from the lowest measured up to about $7 \times 10^{15}$ cm$^{-3}$, the range of doping concentration over which analyses of the variation of carrier concentration with temperature yielded accurate values for $N_D$ and $N_A$. There was also good agreement in this doping range between the values of $N_D$ and $N_A$ determined from the mobility analyses. For values of $N_D + N_A \gtrsim 7 \times 10^{15}$ cm$^{-3}$, $N_D$ and $N_A$ were estimated from the mobility analysis over the entire temperature range as described previously. To account for the variation in the mobility at 77 K with compensation, the empirical data were presented in terms of the product of the ionized impurity concentration and the screening factor from the Brooks–Herring equation (eqn. (9)) using the experimental carrier concentration at 77 K for $n$ and assuming that $b \gg 1$ and that $n^* = n = N_D - N_A$. The experimental points from these analyses and the best empirical solid curve through these points are shown in Fig. 9, along with other experimental results from the literature.

The broken curve in Fig. 9 is the calculated variation in the mobility at 77 K in the relaxation time approximation obtained by Wolfe *et al.*[29] for two different compensation ratios $N_A/N_D$ of 0 and 0.5. When the calculated mobility values are plotted in terms of the product of the total ionized impurity concentration and the Brooks–Herring screening factor, the two different calculated curves in Fig. 3 of ref. 26 both fall on the broken curve of Fig. 9.

Rode has also calculated the mobility at 77 K as a function of free carrier concentration for different values of $(N_D^+ + N_A^-)/n$, the ratio of the total ionized impurity concentration to the free carrier concentration[31, 33], and when these results are plotted in terms of the abscissa of Fig. 9 they also follow the broken curve for values of $(N_D^+ + N_A^-)/n$ from 1 to 10. This agreement indicates that the Brooks–Herring screening term accurately accounts for the variation in the mobility at 77 K with compensation ratio for a fixed carrier concentration.

Kranzer and Eberharter[32] have also calculated the variation in the mobility at 77 K with total ionized impurity concentration, although the compensation ratio or relative values of $N_A$ and $N_D$ used for this calculation were not specified. However, the three inverted triangular points in Fig. 9 are from their calculations for three specific samples for which the compensation ratios were determined, and these three points agree very well with the empirical curve in Fig. 9.

The largest errors in the values of $N_D$ and $N_A$ determined from Hall constant and resistivity measurements at 77 K, using Fig. 9, result from experimental errors, the temperature variation of the scattering factor $r_H$ and the influence of other scattering mechanisms at higher values of $N_D$ and $N_A$ on the initial measure-

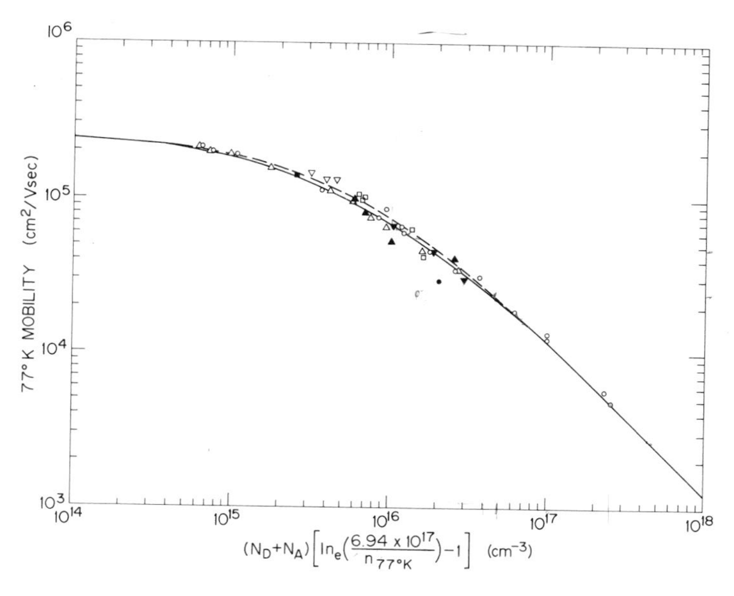


Fig. 9. Empirical (broken) and calculated (solid) curves relating the total impurity concentration in GaAs to the mobility at 77 K. The solid curve is from the calculations of Wolfe *et al.*[29] The data points ○ and △ were obtained from mobility and carrier concentration analyses of data obtained in a magnetic field of 5 kG. The other points are from similar data from the literature: □ Bolger *et al.*[10, 11]; ▽ Maruyama *et al.*[13]; ● Carballés *et al.*[16]; ▲ Kranzer and Eberharter[32]; ▲ Dvoryankin *et al.*[39]; ■ Akasaki and Hara[40].

ments and analyses used to establish the empirical curve. The scattering factor $r_H$ for ionized impurity scattering in the low magnetic field limit can vary from 1.29 at high concentrations and temperatures to 1.60 at low concentrations and temperatures. For a magnetic field of 5 kG the assumption that $r_H = 1$ results in a smaller underestimate of the carrier concentration. The amount of the error depends on the impurity concentration through the mobility, and is estimated to be less than 20% at high impurity concentrations and less than 10% at low concentrations. The Hall coefficient factor for the different scattering mechanisms acting simultaneously has been calculated by Fletcher and Butcher[30], and their calculation at a magnetic field of 5 kG indicates that the assumption that $r_H = 1$ results in an underestimate of the carrier concentration of less than 5%. Calculations of both the drift mobility and Hall mobility for all the important scattering mechanisms have also been given by Rode[33]. Thus the errors arising from using $r_H = 1$ (*i.e.* equal Hall and drift mobility) are probably less than the errors due to small inhomogeneities and errors in sample dimensions for typical samples. Although only measurements of mobility and carrier concentration at 77 K are required to determine $N_D$ and $N_A$ using Fig. 9, the room temperature values should also be examined to help evaluate the quality of a sample, since the room temperature measurements will often give indications of sample inhomogeneity which are not detected in a single measurement at 77 K.

Many different types of inhomogeneity can have an effect on the measured mobility of semiconductors[4, 34] and therefore on the donor and acceptor concentrations determined from mobility measurements. Most of the inhomogeneities considered cause an anomalously low mobility, either as a result of the averaging inherent in the Hall constant and resistivity measurements or as a result of additional carrier scattering. These types of inhomogeneity lead to an overestimate of the total impurity concentration and in this way give an indication of a degradation of sample quality. It has recently been shown, however, that a simple conducting inhomogeneity can lead to an anomalously high mobility determined from Hall effect and resistivity measurements[35, 36]. By introducing conducting homogeneities in high purity epitaxial GaAs, it was possible to increase the measured mobility at 300 K from 7400 $cm^2\ V^{-1}\ sec^{-1}$ to 24 000 $cm^2\ V^{-1}\ sec^{-1}$ and the mobility measured at 77 K from 150 000 $cm^2\ V^{-1}\ sec^{-1}$ to 740 000 $cm^2\ V^{-1}\ sec^{-1}$. Samples with this type of conducting inhomogeneity can appear to be quite uncompensated when this is not actually the case, so the measurement of a high mobility alone is not a sufficient indication of sample quality. The homogeneity of the material must also be established, and this can be done by checking for agreement between the calculated and experimental variation in mobility with temperature, by looking for variations in the resistivity ratio with temperature for van der Pauw measurements or in measurements on different sets of contacts for standard Hall samples, and by examining the magnetic field dependences of the Hall constant[36–38]. Although mobility measurements for routine analysis of sample quality remain very useful, it is clear that the quality or purity of a sample cannot be established conclusively solely by a high mobility at 77 K, particularly when the room temperature value exceeds the theoretical value of about $8000 \pm 1000$ $cm^2\ V^{-1}\ sec^{-1}$, or when room temperature measurements cannot be made.

## FURTHER COMPLICATIONS AND CONCLUSIONS

In addition to some of the problems involved in determining $N_D$ and $N_A$ from electrical measurements as described, there are still further complications which have not been discussed. For example, the conduction band in GaAs is slightly non-parabolic and this should be included in the model for calculating the free carrier concentration, as well as in the mobility calculations[31,33]. The non-parabolicity can be included simply in the model, and it results in a temperature-dependent density-of-states effective mass. However, the temperature dependence of $r_H$ can easily obscure the effect of the non-parabolicity and it is probably sufficiently accurate to use an average effective mass ($m_D^* \approx 0.072\ m_0$) instead of the effective mass at the bottom of the conduction band ($m_0^* = 0.0665\ m_0$)[19]. The temperature variation in $r_H$ also makes the determination of the ratio $N_c/g_1$ in eqn. (2) or eqn. (6), from the intercept of a plot of these equations on semi-log paper, completely unreliable[14-16].

The model given for the impurity excited states is only correct for zero magnetic field, while most practical measurements are made at a field of about 5 kG in order to improve the approximation $r_H \approx 1$. In a magnetic field, the first excited state is split into the 2p, $m = \pm 1$, 0, and 2s states[12], so this should be taken into account in eqn. (6). However, again because of the uncertainty in the actual value and temperature dependence of $r_H$, and because at 5 kG the maximum splitting of these states is only about 1 meV, the neglect of this complication introduces little additional uncertainty into the values of $N_D$ and $N_A$ determined from analysis of Hall coefficient *versus* temperature data.

In conclusion, for GaAs it is possible to make routine determinations of $N_D$ and $N_A$ from a single measurement of the Hall constant and resistivity at a temperature of 77 K by using the data of Fig. 9. The uncertainties in the values determined are mainly due to the sample inhomogeneity and the failure of the assumption $r_H = 1$. However, with careful evaluation of sample inhomogeneity, *e.g.* from the temperature and magnetic field dependence of the resistance ratios in van der Pauw measurements or the magnetic field dependence of $r_H$, the values of $N_D$ and $N_A$ can be determined to better than $\pm 20\%$. If better accuracy than this is required, Hall coefficient measurements over the entire range must be analyzed and evaluated using the calculated mobility including all the relevant scattering mechanisms.

## ACKNOWLEDGMENT

This work was sponsored by the Defense Advanced Research Projects Agency and by the Department of the Air Force.

## REFERENCES

1 O. Lindberg, *Proc. IRE, 40* (1952) 1414.
2 D. J. Oliver, *Phys. Rev., 127* (1962) 1045.
3 R. A. Reynolds, *Solid-State Electron., 11* (1968) 385.
4 A. C. Beer, *Galvanomagnetic Effects in Semiconductors*, Academic Press, New York, 1963, p. 308.
5 G. E. Stillman, C. M. Wolfe and J. O. Dimmock, *J. Phys. Chem. Solids, 31* (1970) 1199.

6 B. F. Lewis and E. H. Sondheimer, *Proc. R. Soc. London, Ser. A*, *227* (1954) 241.
7 H. Ehrenreich, *J. Phys. Chem. Solids*, *9* (1959) 129.
8 S. S. Devlin, in M. Aven and J. S. Prener (eds.), *Physics and Chemistry of II–VI Compounds*, North-Holland Publ. Co., Amsterdam, 1967, p. 551.
9 J. S. Blakemore, *Semiconductor Statistics*, Pergamon Press, New York, 1962, p. 138.
10 D. E. Bolger, J. Franks, J. Gordon and J. Whitaker, *Proc. Int. Symp. on GaAs*, Inst. Phys. and Phys. Soc., London, 1967, p. 16.
11 J. Whitaker and D. E. Bolger, *Solid State Commun.*, *4* (1966) 181.
12 G. E. Stillman, C. M. Wolfe and J. O. Dimmock, in E. M. Pell (ed.), *Proc. 3rd Photoconductivity Conf.*, Pergamon Press, New York, 1971, p. 265.
13 M. Maruyama, S. Kikuchi and O. Mizuno, *J. Electrochem. Soc.*, *116* (1969) 413.
14 D. V. Eddolls, *Phys. Status Solidi*, *17* (1966) 67.
15 D. V. Eddolls, J. R. Knight and B. L. H. Wilson, *Proc. Int. Symp. on GaAs*, Inst. Phys. and Phys. Soc., London, 1967, p. 3.
16 J. C. Carballés, D. Diguet and J. Lebailly, *Proc. 2nd Int. Symp. on GaAs*, Inst. Phys. and Phys. Soc., London, 1969, p. 28.
17 K. S. Shifrin, *Zh. Tekh. Fiz.*, *14* (1944) 43.
18 P. T. Landsberg, *Proc. Phys. Soc., London, Sect. B*, *69* (1956) 1056.
19 G. E. Stillman, D. M. Larsen, C. M. Wolfe and R. C. Brandt, *Solid State Commun.*, *9* (1971) 2245.
20 G. E. Stillman, C. M. Wolfe and D. M. Korn, *Proc. 11th Int. Conf. on the Physics of Semiconductors, Warsaw*, Polish Scientific Publishers, Warsaw, 1972, p. 863.
21 G. E. Stillman, unpublished, 1972.
22 K. Shifrin, *J. Physics (Moscow)*, *8* (1944) 242.
23 H. Brooks, in L. Marton (ed.), *Advances in Electronics and Electron Physics*, Academic Press, New York, 1955, p. 158.
24 C. M. Wolfe, G. E. Stillman and J. O. Dimmock, *J. Appl. Phys.*, *41* (1970) 504.
25 H. Ehrenreich, *J. Appl. Phys.*, *32* (1961) 2155.
26 H. J. G. Meijer and D. Polder, *Physica*, *19* (1953) 255.
27 W. A. Harrison, *Phys. Rev.*, *101* (1956) 903.
28 C. Erginsoy, *Phys. Rev.*, *79* (1950) 1013.
29 C. M. Wolfe, G. E. Stillman and W. T. Lindley, *J. Appl. Phys.*, *41* (1970) 3088.
30 K. Fletcher and P. N. Butcher, *J. Phys. C*, *5* (1972) 212.
31 D. L. Rode and S. Knight, *Phys. Rev.*, *133* (1971) 2534.
32 D. Kranzer and G. Eberharter, *Phys. Status Solidi A*, *8* (1971) K89.
33 D. L. Rode, in R. K. Willardson and A. C. Beer (eds.), *Semiconductors and Semimetals*, Vol. 10, Academic Press, New York, 1975, p. 1.
34 R. T. Bate, in R. K. Willardson and A. C. Beer (eds.), *Semiconductors and Semimetals*, Vol. 4, Academic Press, New York, 1968, p. 459.
35 C. M. Wolfe and G. E. Stillman, *Appl. Phys. Lett.*, *18* (1971) 205.
36 C. M. Wolfe, G. E. Stillman and J. A. Rossi, *J. Electrochem. Soc.*, *119* (1972) 250.
37 C. M. Wolfe, G. E. Stillman, D. L. Spears, D. E. Hill and F. V. Williams, *J. Appl. Phys.*, *44* (1973) 732.
38 C. M. Wolfe and G. E. Stillman, in R. K. Willardson and A. C. Beer (eds.), *Semiconductors and Semimetals*, Vol. 10, Academic Press, New York, 1975, p. 175.
39 V. F. Dvoryankin, O. V. Emel'yanenko, D. N. Nasledov, D. D. Nedeoglo and A. A. Telegin, *Sov. Phys. Semicond.*, *5* (1972) 1636.
40 I. Akasaki and T. Hara, *Proc. 9th Int. Conf. on the Physics of Semiconductors, Moscow*, Nauka, Leningrad, 1968, p. 787.

*Thin Solid Films, 31* (1976) 89–122



# SURFACE AND THIN FILM ANALYSIS OF SEMICONDUCTOR MATERIALS

RICHARD E. HONIG
*RCA Laboratories, Princeton, N.J. 08540 (U.S.A.)*
(Received April 2, 1975; accepted June 17, 1975)

In the field of semiconductor materials characterization, interest has shifted in recent years from bulk problems to surfaces and thin films, and from average composition to three-dimensional concentration profiles. This paper reviews nine major methods available today for the determination of surface and thin film composition, placing particular emphasis on basic principles and up-to-date instrumentation. Major considerations that determine the choice of a given method include: area and depth to be sampled; sensitivity, resolution and reproducibility; number of elements to be detected; and equipment price. To illustrate the capabilities and limitations of these methods, a number of practical examples are presented. Finally, the nine major methods are compared in terms of the most important practical parameters.

## 1. INTRODUCTION

In the field of semiconductor materials characterization, interest has shifted in recent years from bulk problems to surfaces and thin films, and from average composition to three-dimensional concentration profiles. It is the purpose of this paper to review and compare the major methods available for the determination of surface and thin film composition, and to present some practical examples which compare the capabilities and limitations of these methods. In this context, we define "surface" as the outermost atomic layer bounding the solid, while the thickness of a "thin film" is presumed to lie in the range 0.1–100 μm. Because there exists such a bewildering array of modern instrumentation, most of it developed only in recent years, it is desirable first to list each method, and then to present briefly the physical processes involved. The listing is done succinctly in matrix form[1], as shown in Table I. In it, rows represent different types of primary excitation, while columns indicate detected emission. Those methods generally available are presented in capital letters, with surface and thin film instrumentation shown in boxes. For the sake of completeness, Table I also presents (bracketed and in lower-case letters) several specialized methods that are of considerable interest in surface and thin film analysis but for which instrumentation is not yet commercially available; for this reason they will not be included in this review.

TABLE I

SURVEY OF MAJOR METHODS OF MATERIALS CHARACTERIZATION

| PRIMARY EXCITATION \ DETECTED EMISSION | | OPTICAL | X-RAYS | ELECTRONS | IONS (+ AND −) |
|---|---|---|---|---|---|
| PHOTONS | OPTICAL | "AA": ATOMIC ABSORPTION<br>"IR": INFRARED } SPEC-<br>VISIBLE } TROS-<br>"UV": ULTRAVIOLET } COPY | | "ESCA": ELECTRON SPECTR. F. CHEMICAL ANALYSIS<br>"UPS": VAC. UV PHOTOELECTRON SPECTROSCOPY - OUTER SHELL | |
| | X-RAYS | | X-RAY FLUORESCENCE SPECTROMETRY<br>X-RAY DIFFRACTION | "XPS": X-RAY PHOTOELECTRON SPECTROSCOPY - INNER SHELL | |
| ELECTRONS | | | "EPM": ELECTRON-PROBE MICRO-ANALYSIS | "AES": AUGER ELECTRON SPECTROSCOPY<br>"SAM": SCANNING AUGER MICROANALYSIS<br>"SEM": SCANNING ELECTRON MICROSCOPY<br>"TEM": TRANSMISSION ELEC-TRON MICROSCOPY | |
| IONS (+ AND −) | | ["SCANIIR": Surf. Comp. by Anal. of Neutral and Ion Impact Radiation] | [Ion-Induced X-Rays] | | "SIMS": SECONDARY ION MASS SPECTRO METRY<br>"IPM": ION-PROBE MICRO-ANALYSIS<br>"ISS": ION SCATTERING SPECTROMETRY<br>["RBS": Rutherford Backscat-tering Spectrometry] |
| RADIATION | | "ES": EMISSION SPECTROSCOPY | | | "SSMS": SPARK SOURCE MASS SPECTRO-GRAPHY |

There are a number of major considerations that determine the choice of an instrumental method to solve a specific problem in surface or thin film analysis. These considerations include area and depth to be sampled, sensitivity and reproducibility, number of detectable elements and, last but not least, equipment price. Table II surveys nine major methods from this viewpoint, to give the reader a general understanding of what method may be applicable to a given situation.

In addition, there are many pertinent interrelated specific parameters that need to be considered in some detail. They include: the effect of the matrix on impurity sensitivity ("matrix effect"); geometrical effects; charge-up problems with insulating samples, and the field-induced motion of ionized impurities; beam-induced chemical changes in the sample; in-depth concentration profiles and resolution (crater size and shape *versus* detected area); lateral concentration profiles and resolution (primary beam size, rastering and line and area scans); elemental identification (mass resolution); depth scale (sputtering rates and preferential sputtering); lattice damage within the primary ion penetration depth; escape or information depth of the detected emission; the sample consumed during analysis; the time spent on the complete analysis; and surface particle *versus* sputtered species detection.

The following sections will first present in some detail these basic considerations, placing special emphasis on the physical processes involved. This will be followed by a description of instrumentation. Finally, practical

TABLE II

SURVEY OF METHODS FOR SURFACE AND THIN FILM ANALYSIS

| METHOD | PROBE DIAMETER µm | SAMPLING DEPTH µm | SAMPLING DEPTH ATOMIC LAYERS | OPTIMUM DETECTION SENSITIVITY (ppm atomic) | REPRODUCIBILITY (%) | COVERAGE OF ELEMENTS | SPECIAL FEATURES | APPROX. PRICE (K$) |
|---|---|---|---|---|---|---|---|---|
| X-RAY FLUORESCENCE SPECTROMETRY | $10^4$ | 3-100 | $10^4$-$3\cdot10^5$ | 1-100 | ±1 | NEARLY COMPLETE ($Z \geqslant 9$) | QUANTITATIVE; NONDESTRUCTIVE; INSULATORS | 25 |
| ELECTRON-PROBE MICROANALYSIS | 1 | 0.03-1 | $10^2$-$3\cdot10^3$ | 100-1000 | ±2 | COMPLETE ($Z \geqslant 4$) | QUANTITATIVE; "NONDESTRUCTIVE" | 100 |
| SOLIDS MASS SPECTROGRAPHY | 10-100 | 1-10 | $3\cdot10^3$-$3\cdot10^4$ | 0.01-10 | ±20<br>±2 | NEARLY COMPLETE | SEMI-QUANTITATIVE; ION-SENSITIVE PLATES<br>ELECTRICAL READOUT | 100<br>150 |
| ION SCATTERING SPECTROMETRY | $10^3$ | | 1 | 0.1-1% | ±20 | NEARLY COMPLETE NO H, He | SEMI-QUANTITATIVE; IN-DEPTH CONC. PROF. INSULATORS | 40 |
| SECONDARY ION MASS SPECTROMETRY | $10^3$ | | 3 | 0.1-100 | ±2 | NEARLY COMPLETE | SEMI-QUANTITATIVE; IN-DEPTH CONCENTRATION PROFILE | 25-100 |
| ION-PROBE MICROANALYSIS | 1-300 | | 10-1000 | 0.1-100 | ±2 | NEARLY COMPLETE | SEMI-QUANTITATIVE; THREE-DIMENSIONAL CONC. PROFILE | 300 |
| AUGER ELECTRON SPECTROMETRY | 25-100 | | 2-10 | 0.01-0.1%<br>100-1000 | ±20 | NEARLY COMPLETE NO H, He | SEMI-QUANTITATIVE; THREE-DIMENSIONAL CONC. PROFILE | 55 |
| "SAM": SCANNING AUGER MICROANALYSIS | 4-15 | | 2-10 | 0.1-1%<br>1000-10000 | ±20 | NEARLY COMPLETE NO H, He | SEMI-QUANTITATIVE; THREE-DIMENSIONAL CONC. PROFILE; TWO DIM'L AUGER IMAGES | 100 |
| "XPS": X-RAY PHOTOELECTRON SPECTROSCOPY ("ESCA") | $10^4$ | | 2-10 | 0.1-1%<br>1000-10000 | ±20 | NEARLY COMPLETE NO H, He | SEMI-QUANTITATIVE; VALENCE STATES | 150 |

examples will be presented that illustrate the capabilities and limitations of each method, and permit a general comparison to be made.

## 2. BASIC CONSIDERATIONS

### *2.1. Interaction of primary excitation with matter*

#### *2.1.1. X-rays*

When X-rays enter a solid sample, many interaction phenomena occur of which two are of interest to the present discussion: the production of secondary X-rays, utilized in X-ray fluorescence spectroscopy (XRFS), and photoelectron emission. The latter is employed in X-ray photoelectron spectroscopy (XPS), sometimes also referred to as electron spectroscopy for chemical analysis (ESCA).

*2.1.1.1. X-ray fluorescence spectroscopy (XRFS).* The process of characteristic secondary X-ray emission is shown schematically in Fig. 1(a) for the case of silicon. A primary X-ray of sufficient energy ionizes an atom, *e.g.* by creating a K shell vacancy, which is then filled, in this example, by an $L_{2,3}$ electron in accordance with certain selection rules. The transition produces a Si $K\alpha_{1,2}$ X-ray with energy $E(Z) = E_K(Z) - E_{L_{2,3}}(Z)$, conforming with the energy level diagram shown in Fig. 1(a). These characteristic X-rays originate at depths ranging from about 3 µm to 100 µm; thus this method is applicable to the characterization of thin films of micrometer thickness. Because of the great penetration

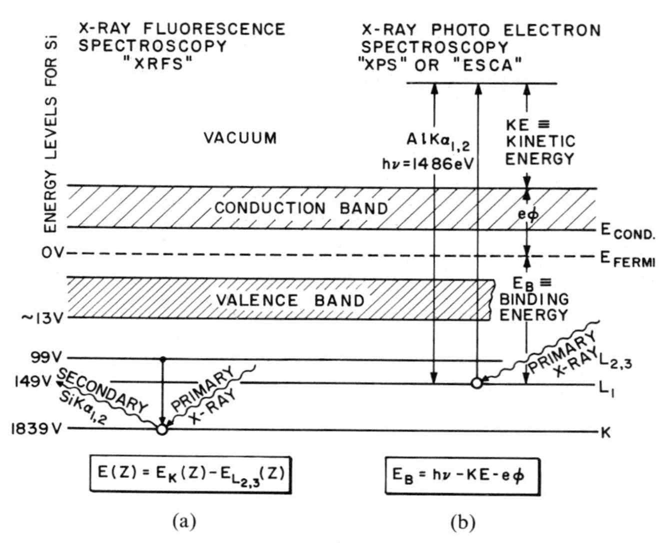


Fig. 1. Schematic representation of X-ray processes for the case of silicon: (a) X-ray fluorescence (secondary X-ray emission); (b) X-ray photoelectron emission.

depth of the primary beam, its physical and chemical effects within the shallow epitaxial layer are negligible, which makes XRFS an essentially non-destructive method, in contrast to most of the other methods to be described later.

*2.1.1.2. X-ray photoelectron spectroscopy (XPS or ESCA).* In the XPS or ESCA process, a primary X-ray of appropriate energy $h\nu$, usually the Al $K\alpha_{1,2}$ line, impinges on the surface of a solid sample in a vacuum, ejecting a photoelectron as indicated in Fig. 1(b). From its measured kinetic energy KE the binding energy $E_B$ is deduced: $E_B = h\nu - \text{KE} - e\phi$, where $e\phi$ is the work function of the surface. The binding energy is characteristic of the element present in the sample, and furthermore provides information concerning its molecular state. These photoelectrons have typical escape depths ranging from about 5 to 30 Å; thus XPS or ESCA characterizes the surface region of a sample in terms of its composition and molecular state. Since physical and chemical effects of the primary beam on the atomic layers near the surface are negligible, XPS or ESCA is an essentially non-destructive method.

*2.1.2. Electrons*

Of the many processes occurring when an electron beam interacts with a solid, two are of special interest to materials characterization: the generation of secondary X-rays, used in electron probe microanalysis (EPM), and the generation of secondary electrons, used in Auger electron spectroscopy (AES) and scanning Auger microanalysis (SAM). Figure 2 shows, in diagrammatic form[2], the interaction of an electron beam with a solid, indicating in particular the active volumes from which secondary X-rays and electrons will be emitted. Typical emission depths range from 0.3 to 1 μm for secondary X-rays and from 5 to 30 Å for Auger electrons, as will be discussed in more detail later.

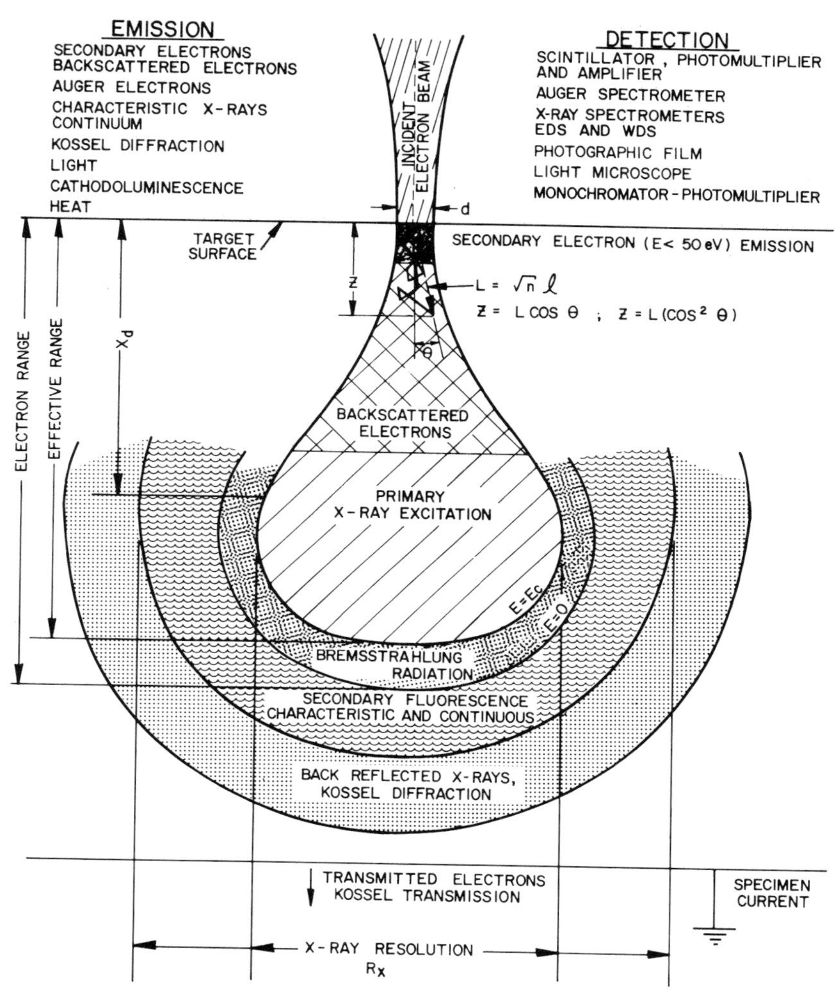


Fig. 2. Schematic representation of the interaction of an electron beam with a solid (after Beaman and Isasi[2]).

*2.1.2.1. Electron probe microanalysis (EPM).* The basic process of characteristic X-ray emission due to primary electron impact is shown diagrammatically for the case of silicon in Fig. 3(a). It differs from XRFS only in terms of the primary excitation, which in this case is an electron with energy sufficient to create a K-shell vacancy, subsequently filled by an electron from an outer shell, such as $L_{2,3}$. Here again selection rules developed by quantum physics allow only selected transitions. An X-ray carries away the available energy $E(Z)$. In this example $E(Z) = E_K(Z) - E_{L_{2,3}}(Z)$, corresponding to the Si $K\alpha_{1,2}$ line. While the primary electron beam (typical energy 5–20 keV) can be focused into a spot less than 1000 Å in diameter, the typical X-ray emitting pear-shaped

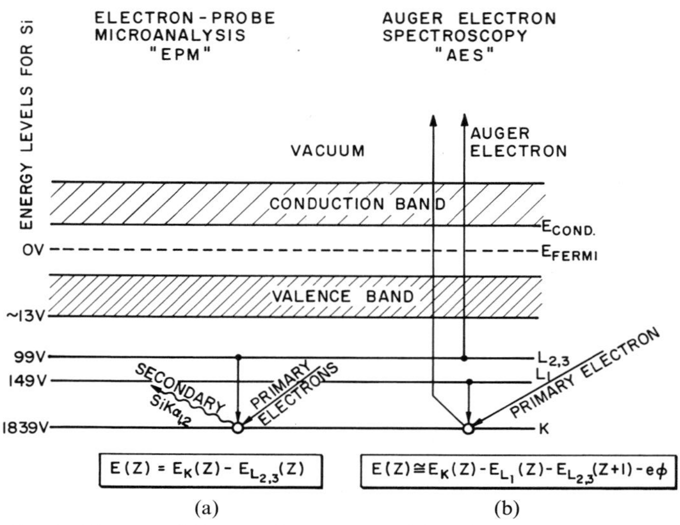


Fig. 3. Diagram of electron interactions with a silicon sample: (a) electron probe microanalysis: electron in, X-ray out; (b) Auger electron spectroscopy: electron in, Auger electron out.

volume shown in Fig. 2 has micrometer dimensions. The primary electron beam can produce chemical effects, such as bombardment-induced desorption, in certain materials but otherwise EPM may be considered a non-destructive method.

*2.1.2.2. Auger electron spectroscopy (AES) and scanning Auger micro-analysis (SAM).* The process of Auger electron emission is indicated schematically in Fig. 3(b). For the example shown, K-shell ionization results from primary electron impact and the K-shell hole is filled by an outer-shell electron ($L_1$ in this instance). The available energy is given to the so-called Auger electron, originating in this case at the $L_{2,3}$ level. Thus, the atom remains in a doubly ionized state, and the entire process would be labeled "$KL_1L_{2,3}$". The net kinetic energy of the Auger electron is given[3] by the approximate expression

$$E(Z) = E_K(Z) - E_{L_1}(Z) - E_{L_{2,3}}(Z+1) - e\phi \tag{1}$$

where $Z$ is the atomic number of the matrix and $e\phi$ is the work function of the surface.

*2.1.3. Ions*

When an energetic ion collides with a solid a number of processes result, as indicated in Fig. 4. A fraction of the ions is backscattered by the surface atoms, mostly in elastic or inelastic binary collisions. The remaining ions penetrate the solid and transfer their energy in a series of collision cascades to the lattice. The energetic recoil atoms initiate secondary and tertiary collision cascades, some of which produce sputtering, *i.e.* the emission into the vacuum of lattice atoms near the surface in either neutral or charged form. If rare gas ions are employed, the elastic scattering process provides energy spectra characteristic of the mass of the scattering centers; thus ion scattering spectrometry (ISS)

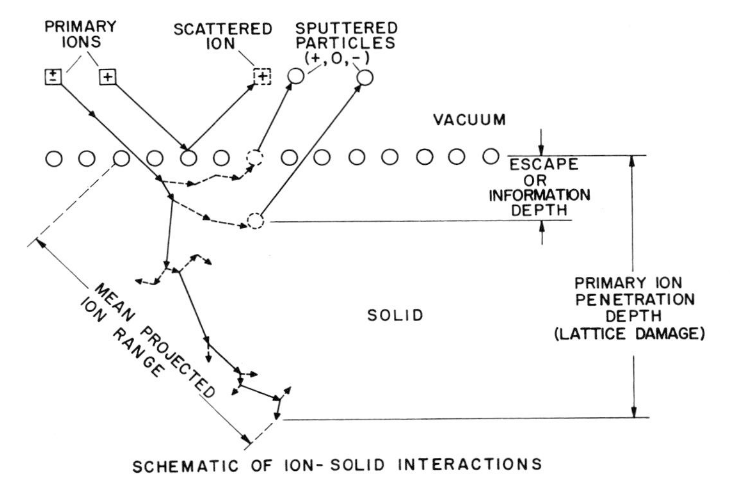


Fig. 4. Schematic diagram of various processes resulting from the collision of a keV ion beam with a solid surface.

can be used to determine the composition of the outermost atomic layer of a solid. Since the sputtering process expels lattice particles, mostly as neutrals but some as ions, from the solid into the vacuum, it can be employed in conjunction with a mass spectrometer to characterize the composition of the solid near the surface. Sputtering continuously uncovers a fresh surface; thus in-depth concentration profiles of major and minor constituents and trace impurities can be obtained down to a depth of several micrometers.

If keV rare gas ions are used, scattering due to elastic binary collisions with surface atoms predominates, and thus the "escape depth" of ions used in ISS is limited to the outermost one or two atomic layers. On the other hand, the penetration depth of primary ions producing sputtering can be considerable, as indicated in Fig. 4, and is a function of primary ion energy and angle of incidence. The escape depth of the sputtered particles is related to the penetration depth[4] and is estimated to range from about 3 to 20 atomic layers, depending on primary ion energy.

*2.1.3.1. The scattering process.* Figure 5 is a schematic representation of the elastic binary collision process for the special case when the primary ion is scattered through a lab angle $\theta = 90°$. From the basic formula, also shown in Fig. 5, the mass of the scattering center can be deduced from the energy of the scattered ion; thus ion scattering spectrometry (ISS) is truly a surface analytical tool. Furthermore, it is also clear from the formula that 90° scattering occurs only if $M_2 > M_1$. The intensity of a specific scattered peak is given as[4]

$$I_{sc}^{+} = kC \left(\frac{d\sigma}{d\Omega}\right) P_i(v_{sc})\, G \tag{2}$$

where $k$ is a proportionality constant, $C$ is the surface concentration of specific scattering centers, $d\sigma/d\Omega$ is the differential scattering cross section per unit

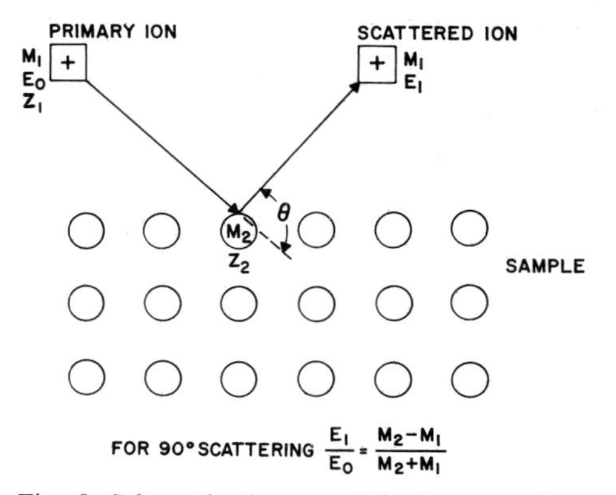


Fig. 5. Schematic diagram of the ion scattering process, and the formula for 90° scattering.

solid angle, in square centimeters, $P_i$ is the probability that a scattered ion remains ionized, which is a function of its scattered velocity $v_{sc}$, and $G$ is a factor taking into account geometric shielding. While scattering cross sections can be computed, not much is yet known about either ionization probability or geometric shielding. Thus scattered intensities cannot be predicted *a priori*, and calibration samples are needed to determine surface concentrations.

*2.1.3.2. The sputtering process.* Sputtering, the emission by ion impact of neutral and charged particles from the surface region of a solid, has been reviewed in detail by many writers, *e.g.* Carter and Colligon[5]. Here we shall discuss only its capability to remove surface layers at a controllable rate, and its analytical application in secondary ion mass spectrometry (SIMS) and ion probe microanalysis (IPM).

The sputtering yield $Y$ and the sputtering rate $\dot{S}$ are two very important parameters since they enter into many considerations, in particular the depth scale of an in-depth analysis. Yield $Y$ is defined as the total number of secondary particles (mainly neutrals) sputtered per primary ion colliding with the solid surface:

$$Y \equiv \frac{N_s/A}{N_p^+/A} = \frac{\dot{N}_s/A}{\dot{N}_p^+/A} \tag{3}$$

where $N$ is the number of particles, $\dot{N}$ is their arrival or departure rate per second, $A$ is the target area in square centimeters, and subscripts p and s refer to primary and secondary, respectively. Sputtering yields are known to be a function of many parameters, in particular: the atomic mass $M_1$ and number $Z_1$, the initial energy $E_0$, and the angle of incidence $\phi$ of the primary ions; the atomic mass $M_2$ and number $Z_2$, and the binding energy (heat of sublimation) of the surface atoms; the crystal structure and orientation of the lattice; and the surface roughness of the sample. Values of $Y$ for various 500 eV ions are known for most elements and some compounds from Wehner's work[6] and from Carter and Colligon's tabulations[5]. These sputtering yield data, valid for normal incidence, have been collected and are presented in graphical form in Fig. 6 for several 500 eV ions. Values for higher energies can be estimated with the help of a scale factor.

The sputtering rate $\dot{S}$, in particles per second per unit area, is derived from the definition of sputtering yield $Y$ as

$$\dot{S} \equiv \frac{\dot{N}_s}{A} = \frac{Y\,\dot{N}_p^+}{A} = \frac{Y\,I_p^+}{1.60\times 10^{-13}\,A} \quad \mathrm{sec}^{-1}\,\mathrm{cm}^{-2} \tag{4}$$

where $I_p^+$ is the primary ion current in microamps. For a uniform primary current density, this definition yields a sputtering rate

$$\dot{S} = 3.75\times 10^{-2}\ Y\bar{V}J_p^+\ \ \text{Å}\ \mathrm{min}^{-1} \tag{5}$$

and for a Gaussian current density distribution

$$\dot{S}_{max} = 3.45\times 10^{-2}\ Y\bar{V}I_p^+/d^2\ \ \text{Å}\ \mathrm{min}^{-1} \tag{6}$$

where $\dot{S}_{max}$ is the maximum rate at the center of a Gaussian crater, $Y$ is the sputtering yield, *i.e.* the number of sputtered particles per primary ion, $\bar{V}$ is the average atomic volume ($\equiv M/N_0\delta$ Å$^3$), $M$ the gram-molecular weight, $N_0$ Avogadro's

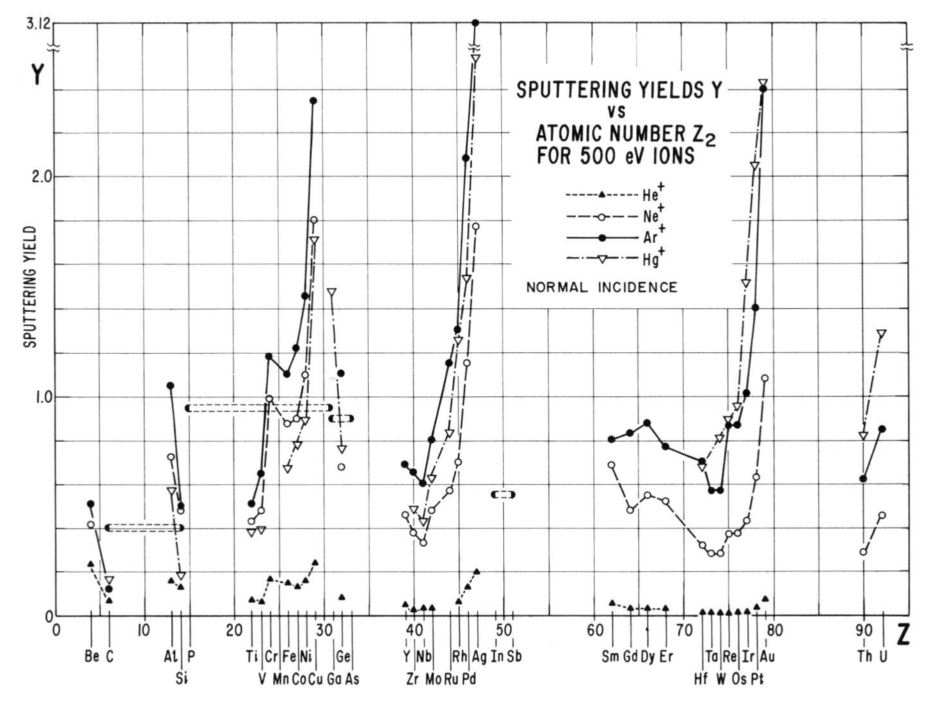


Fig. 6. Sputtering yields $Y$ for 500 eV ions *vs.* atomic number $Z_2$ of the surface atoms. The values given are for normal incidence and are based on Wehner's results[6], the tabulations by Carter and Colligon[5] and the present author's own compilation.

number; $\delta$ the density (g cm$^{-3}$); $J_p^+$ the primary ion current density (µA cm$^{-2}$), $I_p^+$ the primary ion current (µA), and $d$ the beam width at half maximum intensity (FWHM) (cm).

The sputtering yield $Y$ and sputtering rate $\dot{S}$ are well-defined quantities provided the solid is made up exclusively of one kind of atom, which rarely obtains in practice. Thus it is necessary to discuss "preferential sputtering", *i.e.* the effects of individual sputtering rates of a multicomponent system on the composition of the surface and on the overall sputtering rate. Recently, Tarng and Wehner[7] used Auger electron spectroscopy (AES), and Dahlgren and McClanahan[8] used a sputtering technique, to demonstrate the following facts which apply under equilibrium conditions to a two-component system AB for which $Y_A/Y_B$. First, the concentration ratio of surface particles $(C_A/C_B)_{surf}$ is equal to $(Y_B/Y_A)$ $(C_A/C_B)_{bulk}$, *i.e.* at the surface the concentration ratio is increased, compared with the concentration ratio existing in the bulk, by the ratio of individual sputtering yields. Secondly, the overall sputtering rate $\dot{S}_{AB} = \dot{S}_A$, *i.e.* it is limited by and equal to the slower rate. Thirdly, the composition of the vapor phase above the surface accurately represents the bulk composition, *i.e.* $(C_A/C_B)_{vapor} = (C_A/C_B)_{bulk}$.

Two important conclusions may be drawn from these findings. Surface analytical methods, such as ISS and AES, will yield results that reflect the actual concentration of components at or near the surface at a given moment; these concentrations may differ significantly from the underlying bulk values. It should be noted here that for ISS the information depth is limited to the outermost atomic layer or two, while in the case of AES the "escape depth" of Auger electrons ranges from about two to six atomic layers[9]. On the other hand, methods based on the analysis of sputtered particles in the vapor phase, such as secondary ion mass spectrometry (SIMS), will yield concentrations typical of the underlying atomic layers contained within the information depth.

In this context it is appropriate to review in more detail the different depth concepts already indicated in Fig. 4. The bombarding ion has a mean projected range in the solid which is a function of its primary energy $E_0$ and of many other variables. From this mean projected range and the angle of incidence, a primary ion penetration depth is readily derived which represents the region of lattice damage. For a 20 keV ion, this quantity is typically of the order of 250 Å, while at 3 keV it is about 40 Å. The secondary ions used to analyze surfaces and thin films are estimated to have an escape or information depth amounting to roughly one-quarter of the penetration depth, *i.e.* about 60 Å at 20 keV and 10 Å at 3 keV primary energy.

The analytical application of the sputtering process to secondary ion mass spectrometry (SIMS) and ion probe microanalysis (IPM) makes it desirable to discuss briefly the process of secondary ion production. Present ideas concerning positive ions, proposed by Slodzian and Hennequin[10], have been reviewed by Castaing and Hennequin[11] and more recently by Evans[12, 13], who consider two basic situations: the "kinetic" process and the "chemical" process. The following is a brief description of their composite views. The "kinetic" process occurs in clean metal and semiconductor samples bombarded by rare gas ions.

As a result of the collision cascades mentioned earlier, lattice particles are emitted into the vacuum, some of them in a metastable excited state from which they can be transformed, by Auger de-excitation, into positive ions near the solid surface. On the other hand, "chemical" ionization is the term applied to the case where chemically reactive species, such as oxygen, are present in the sample and, because of their high electron affinity, reduce the number of free conduction band electrons. This lowers the neutralization probability for secondary ions formed in the solid and permits them to be emitted as positive ions. The reactive species may already be present in the solid, *e.g.* as an oxide[14], or may be introduced into the system either as a low pressure gas[10] or as the primary bombarding ions. Both methods of oxygen introduction were explored by Benninghoven[15], and the latter method has since been described and used extensively by Andersen[16–18]. McHugh has recently reviewed[19] in some detail various proposed mechanisms of secondary ion emission. The above studies, as well as a recent work by Lewis *et al.*[20], show conclusively that the "chemical" mode of positive ion production depends markedly on the presence in the sample of oxygen or other elements with high electron affinity.

To make quantitative measurements by SIMS or IPM, absolute ion yields, *i.e.* the ratios of secondary ion to incident primary ion, must be known. So far, only Benninghoven and coworkers[21, 22] seem to have obtained reliable results for positive secondary ions, measured under carefully controlled conditions for several clean and oxidized metal surfaces. Negative secondary ion yields can be substantially enhanced by bombarding the sample with $Cs^+$ ions, as first established by Krohn[23] and subsequently applied by Andersen[17, 18].

### 2.2. *In-depth concentration profiles and resolution*

The analysis of surfaces is of great practical importance, but the most frequently asked question concerns thin layers: how does the concentration of a given component in a thin layer vary with depth? This variation, generally called the "in-depth profile", is an elusive quantity which is difficult to determine for a number of reasons. The fidelity with which a measured profile depicts an actual concentration variation depends on the "depth resolution" of the method, a term which has frequently been used but not adequately defined in the literature. Figure 7 serves to illustrate the major parameters that underly the definition of this term. If two layers of width $w$ and concentration $C$ of a given minor constituent are built into a lattice at depths $d_1$ and $d_2$, the measured profiles will broaden into the shapes shown at $w_1$ and $w_2$. It is evident that the area under the curves must remain constant, *i.e.* $Cw = A_1 = A_2$. The deformation of these originally rectangular profiles into the shapes shown is due to several effects, in particular to the Gaussian distribution of the sputtering beam density which produces craters with a bell-shaped cross section[24]. This results in secondary particles being emitted from different depths and sometimes different layers, as shown for a specific case in Fig. 8(a). This effect can essentially be eliminated by rastering or defocusing the sputtering beam and accepting secondary ions only from the central flat portion of the large sputtered area shown in Fig. 8(b). In the case of Auger electron spectroscopy, the detecting electron beam is one

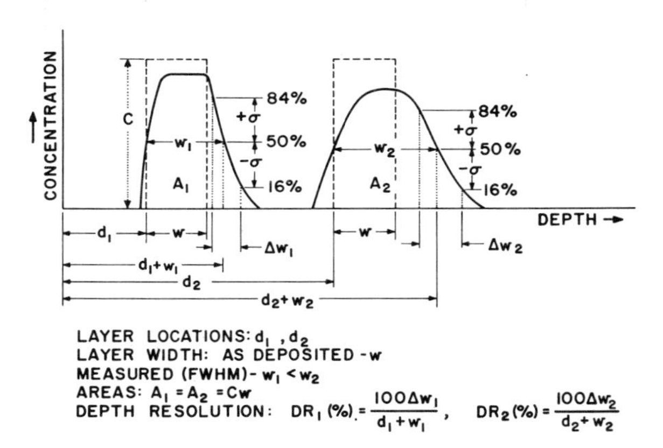


Fig. 7. Major parameters for in-depth concentration profiles.

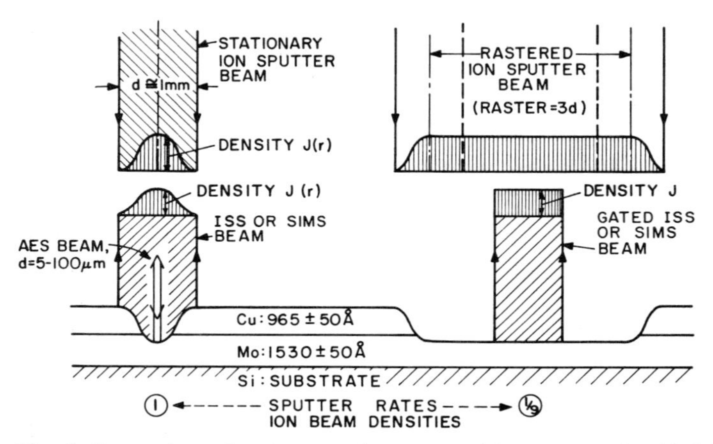


Fig. 8. Comparison of stationary and raster–gated beam geometries. (a) A Guassian current density distribution of the primary ion beam produces a Gaussian-shaped crater and secondary ion current density, resulting in substantial contributions from the crater walls for ISS and SIMS but not for AES. (b) Raster–gating produces a uniform secondary ion density, with the entire detected signal coming from the flat crater bottom. However, sputtering rate as well as secondary ion intensity are reduced by at least one order of magnitude.

to two orders of magnitude smaller than the ion beam used for sputtering, which eliminates the need for rastering, as discussed in more detail later.

A second factor contributing to the deformation of an initially rectangular concentration profile is the information depth of the method employed. A third factor is the lattice damage caused by the primary particles and the ensuing collision cascades. Some of the displaced lattice particles will be driven deeper into the lattice in fresh encounters with new primary particles; thus this effect is cumulative. Both information depth and lattice damage can be minimized by reducing the primary energy $E_0$, but they cannot be eliminated completely since they are characteristic of the basic sputtering process. Referring again to Fig. 7, it is logical to define "depth resolution" in terms of the quality of an in-depth profile of an originally rectangular signal, specifically as the broadening of the trailing edges. The abscissae associated with 84% and 16% of the signal (±one standard deviation from the half maximum value) are chosen to define depth resolution (in percent) as

$$DR_1 = \frac{100\,\Delta w_1}{d_1 + w_1} \qquad DR_2 = \frac{100\,\Delta w_2}{d_2 + w_2} \tag{7}$$

with the quantities as defined in Fig. 7. This convention is frequently used but not fully explained in the literature. As a rule of thumb, it may be stated that under optimum conditions depth resolution may amount to about 3% of the associated depth.

### 2.3. *Sample consumption and sampling depth for SIMS and IPM analyses*

In the analysis of surfaces and thin films by SIMS and IPM, sample consumption obviously plays a major role because it limits the resolution (either lateral or in-depth) and detection sensitivity attainable. Morabito and Lewis[25] point out that a minimum sample volume is required to determine the level of a given impurity to a desired precision. This volume is a function of impurity concentration, precision, instrumental efficiency, the secondary ion/sputtered particle ratio, atomic density and isotopic abundance. As an example, Morabito and Lewis state that a sample volume of 100 $\mu m^3$ is required to detect 10 parts per million atomic (ppma) of mono-isotopic Al with ±3% precision, assuming a secondary ion/sputtered particle ratio of $10^{-3}$. For a primary ion beam diameter of 100 μm, this amounts to a 130 Å layer removed from the sample during the analysis; this thickness is defined as the "sampling depth". McHugh[19] has presented in graphical form the detection sensitivity as a function of primary beam diameter. Assuming a primary beam density of 5 mA $cm^{-2}$ and a 100 sec collection time, corresponding to the removal of a $10^4$ Å layer, he finds a 0.1 ppm detection limit, equivalent to the estimate of Morabito and Lewis just quoted.

### 2.4. *Surface charging problems of insulators*

When insulating samples are to be analyzed by methods that utilize either electrons or ions for the primary beam, the surface potential of the sample will be affected. The value of this potential depends on the arrival rate and polarity of primary particles, their charge leakage rate to ground, and the emission rate and polarity of secondary particles, in other words the net loss or gain of charge density. If the insulator contains impurities that are readily ionizable, such as alkalis, the electrostatic field set up by the surface charges will move the impurity ions towards or away from the surface, depending on polarity. Furthermore, the presence of surface charges will affect the position and effective energy of the primary beam.

To avoid these problems, various schemes have been tried to obtain charge neutralization. In the case of primary beams of positive ions used in ion scattering[26] and secondary ion mass spectrometry, it is possible to neutralize charges with thermal electrons of appropriate density. For the primary electron beams used in Auger electron spectroscopy, it is sometimes possible to balance primary and secondary electron densities by employing grazing incidence[27] for the primary beam. If everything else fails, it is helpful to increase surface conductivity by coating the insulator surface with a conducting layer a few hundred ångströms thick which is eventually sputtered off in the course of the analysis.

## 3. INSTRUMENTATION

### *3.1. X-ray fluorescence (secondary emission) spectrometry*

X-ray fluorescence spectrometry (XRFS), sometimes referred to as X-ray secondary emission spectrometry, is a well-established method for the analysis of films up to 100 μm thick. Since this method has been described in detail by Bertin[28] and Gilfrich[29], it will be reviewed here only briefly. The components of a wavelength-dispersive system are indicated in diagrammatic form in Fig. 9. A beam of primary X-rays irradiates sample A which emits secondary X-rays. After passing through collimator D, the secondary X-rays are diffracted by analyzer crystal E and recorded, after collimation by F, by the detector–amplifier system. To minimize the absorption of low energy radiation ($<1$ keV) in the instrument, the entire instrument (from X-ray tube window to detector window) is evacuated. This allows all elements down to and including fluorine ($Z = 9$) to be recorded. Detection sensitivities range from about 100 ppma for the lighter elements to 1 ppma for the heavier ones, with an optimum reproducibility of $\pm 1\%$. The major advantages of this method are its rapidity and the fact that samples can be analyzed in a non-destructive fashion. In recent years, the use of energy-dispersive systems for XRFS has been explored. Such systems consist of a solid state Si(Li) detector and a multichannel pulse-height analyzer. Even though its resolving power is limited, this method is useful as a fast qualitative survey tool.

### *3.2. Electron probe microanalysis*

Electron probe microanalysis (EPM) is a well-established method for the elemental analysis of sample volumes of micrometer dimensions. Since it has been fully described by a number of authors[2, 30–33], its capabilities need to be reviewed only briefly. Figure 10 shows diagrammatically the major components of an electron probe microanalyzer, with particular emphasis on the associated circuitry which permits the instrument to produce:

(1) complete characteristic X-ray spectra of a fixed point less than 1 μm in diameter, yielding quantitative answers;

(2) line scans in either the $x$ or the $y$ direction over a maximum length of about 500 μm, giving quantitative X-ray information for one constituent per spectrometer; on angle-lapped specimens a line scan produces depth profiles of micrometer resolution;

(3) area scans, over a maximum area of about 500 μm square, providing qualitative X-ray information concerning one constituent at a time; and

(4) scattering electrons, providing topographical information equivalent to that obtained in a scanning electron microscope.

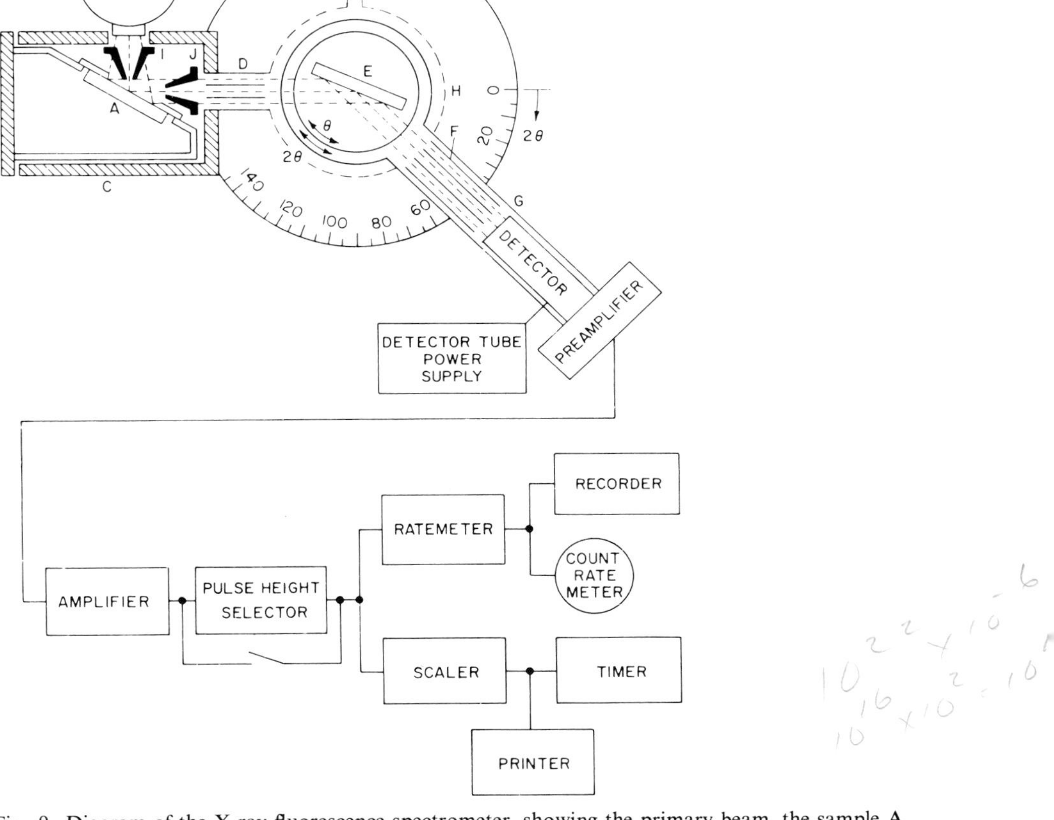


Fig. 9. Diagram of the X-ray fluorescence spectrometer, showing the primary beam, the sample A, the collimated secondary beam, the analyzer crystal E, the diffracted beam lined up by the collimator F, the detector and associated supplies. For the detection of light elements, the entire housing is evacuated (after Bertin[28]).

EPM combines adequate detection sensitivity (about 100 ppma) with good reproducibility (±2%). It is applicable to all elements down to Be ($Z = 4$), is basically non-destructive, and can yield absolute concentrations in the case of binary compounds. To obtain the composition of more complex samples, known standards are required for comparison.

*3.3. Spark source mass spectrometry*

This well-established technique has been applied for many years to the survey analysis of bulk materials, but more recently modifications in the source

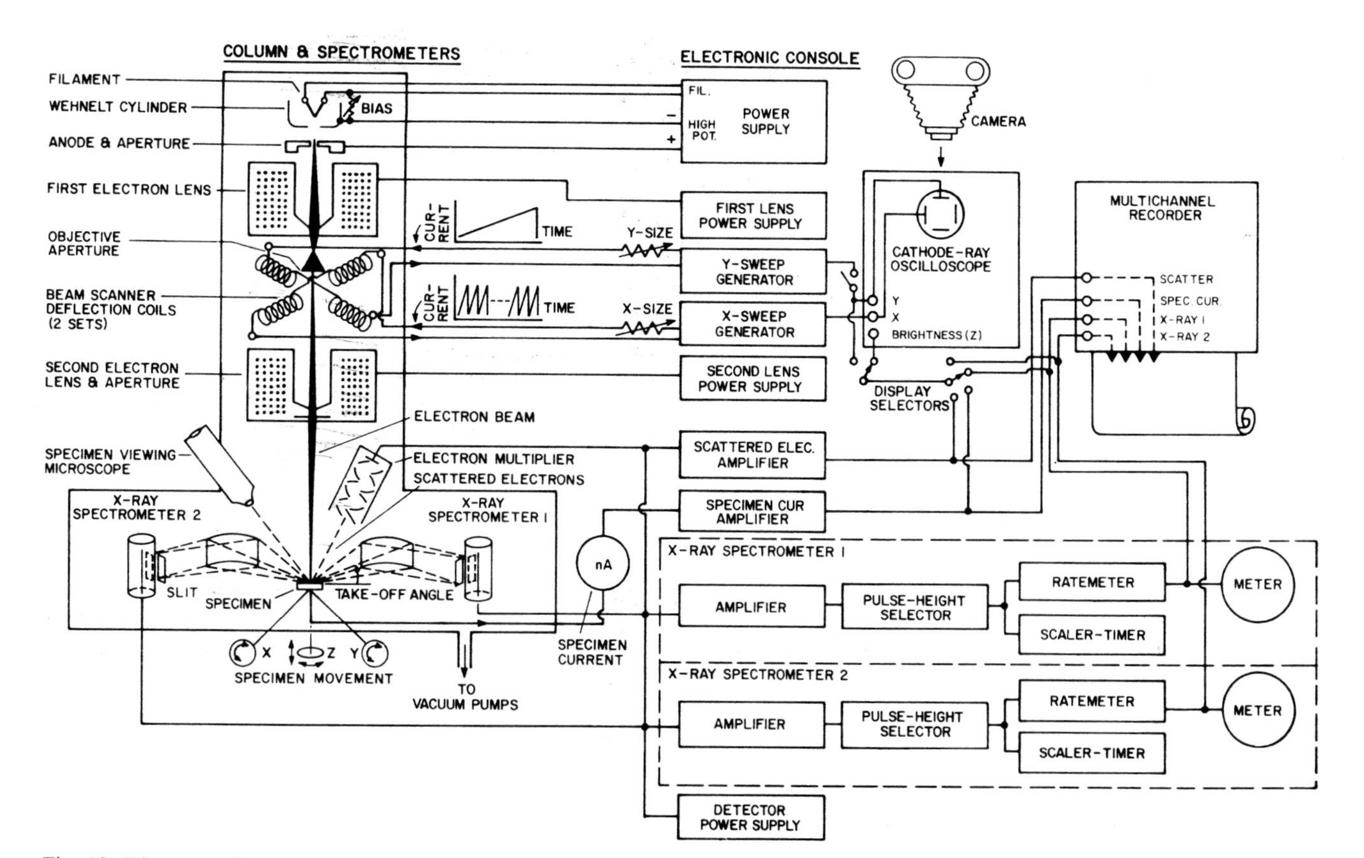


Fig. 10. Diagrammatic representation of the electron microprobe, showing electron-optical column, X-ray spectrometers and associated electronic supplies and read-out systems (after Bertin[30]).

design have extended its use to the analysis of films of micrometer thickness[34]. The basic components of a typical double-focusing spark source mass spectrograph (SSMS) are shown in the diagram of Fig. 11. They include the source region where the solid sample is vaporized, ionized and accelerated; an electrostatic analyzer for energy selection; a magnetic analyzer for mass selection; and an ion-sensitive plate for the simultaneous detection of the mass-dispersed ions over a wide range, typically from $M = 8$ to $M = 250$. Alternatively, the plate can be replaced by a sequential electrical detection and read-out system. The application to thin- film analysis has been made possible by the development of an automatic spark gap control for the source electrodes[35] and surface scanning unit[36]. This system permits plane films, several square centimeters in extent, to be analyzed under constant and reproducible experimental conditions, at sampling depths that range from 0.3 to 10 μm depending on the spark parameters.

SSMS is a fairly fast semi-quantitative survey technique which will detect all elements with high sensitivity, usually to one ppma or better. It can be made quantitative through the use of an electrical detection system.

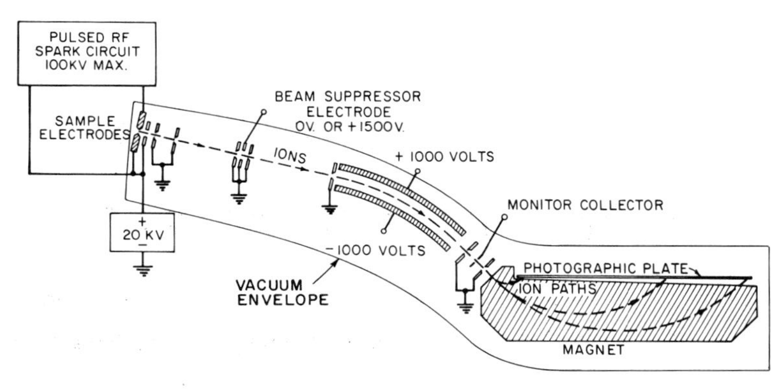


Fig. 11. Spark source mass spectrograph used for the analysis of solid films. The diagram shows the source region, the electrostatic and magnetic analyzers and the ion detector plate.

### *3.4. Ion scattering spectrometry*

An ion scattering spectrometer (ISS) of recent design, commercially available from the 3M Company[26], is schematically shown in Fig. 12. Its major components include a primary ion source with small energy spread but without mass selector, a target holder and manipulator, a charge neutralization filament, a 127° cylindrical electrostatic energy analyzer (ESA) and a channel electron multiplier. All these components are mounted in an ultrahigh vacuum system. The primary beam of rare gas ions (initial energy range 1–3 keV) is focused into a 1 mm diameter spot on the target, giving rise to a current density of about 10 μA $cm^{-2}$. Since the sample surface makes an angle of 45° with the nearly circular primary beam, the sputtered hole is elliptical in shape. The cylindrical ESA accepts a band of ions scattered through 90° which includes those coming from regions near the crater wall, a fact which limits depth resolution. To eliminate this problem it is necessary

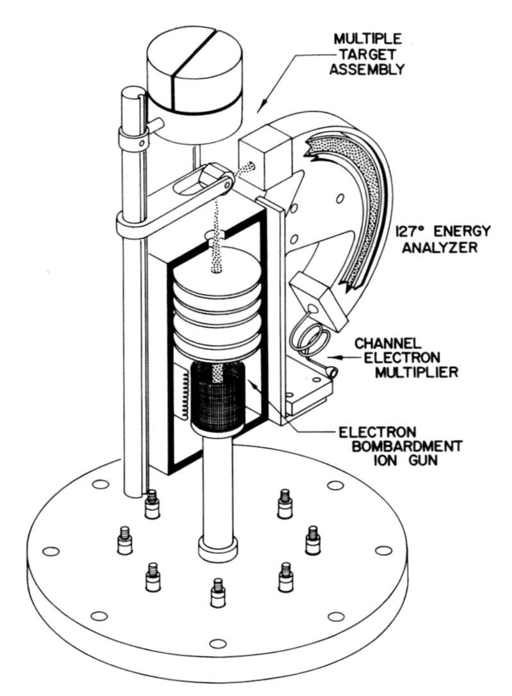


Fig. 12. Schematic diagram of the 3M ion scattering spectrometer (after Goff and Smith[26]).

to make the sputtered crater large with respect to the detected area. This can be accomplished by rastering the primary beam, at the expense of the sputtering rate, or else by shortening the entrance slit of the ESA, which reduces the scattered intensity and thereby the detection sensitivity.

The energy spectrum obtained by ISS yields limited dispersion for higher masses which at times makes positive identification of a specific component difficult or even impossible. Recently, Rusch *et al.*[38] have described a dual purpose instrument that permits scattered primary ions to be recorded at the same time as sputtered secondary ions. This is done by attaching to the 3M ISS instrument described above a secondary ion mass analyzer (SIMS) of the quadrupole type. The latter will be described later.

### *3.5. Ion probe microanalysis*

As discussed by Evans[12], ion probe microanalyzers (IPM) are conveniently grouped in two classes: (1) the direct-imaging analyzer developed by Castaing and Slodzian[14], and (2) the scanning ion microprobe designed by Liebl[39]. Since Evans has discussed in considerable detail the outstanding characteristics of these two types of instruments, the reader is referred in particular to Table I in his review[12] for a summary of their essential features. In the present paper the discussion will be limited to a brief outline of basic operating principles and major components.

The direct-imaging mass analyzer, in its latest version[25], is commercially available from Cameca. As indicated in Fig. 13, primary positive or negative ions are produced in a duoplasmatron and are focused, without mass analysis, into a target spot typically 25–200 μm in diameter. Flat-bottomed craters are obtained by rastering the primary beam. Secondary positive or negative ions are extracted from an area approximately 200 μm × 200 μm, imaged by an immersion lens, and mass- and energy-analyzed in the prism–mirror–prism system. The mass-resolved image is accelerated, converted into electrons, and detected by various means. For quantitative electrical measurements, and to ensure adequate depth resolution, a mechanical aperture is placed in the secondary image plane. The analyzer has a mass resolution of up to 1000, and an optimum lateral imaging resolution of about 0.8 μm. Since the direct-imaging analyzer records all image information simultaneously, it yields the desired information in much less time than the sequential system of the ion microprobe to be discussed next.

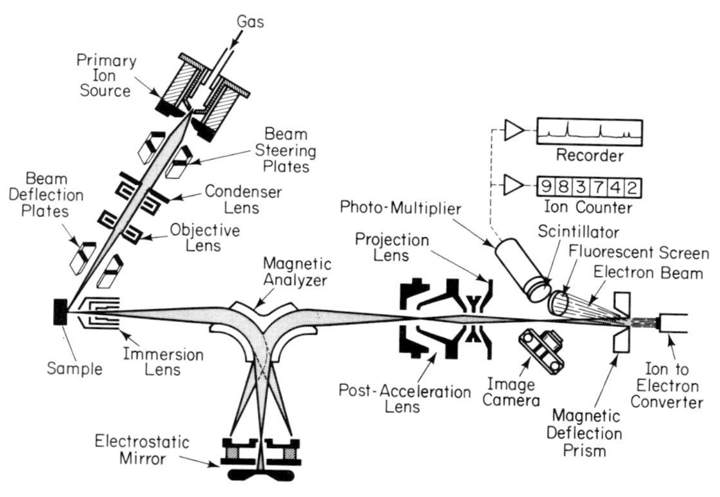


Fig. 13. Schematic diagram of the Cameca direct-imaging mass analyzer (after Evans[12]).

A commercial scanning ion microprobe, available from ARL[41] and shown in Fig. 14, is based on Liebl's[39] design. It uses a duoplasmatron ion source in combination with a mass selector and focuses the primary beam into a 2 μm spot which can be rastered over the target. The secondary ions are extracted, analyzed in a double-focusing mass spectrometer, and detected and imaged by various means. Image information, recorded by rastering, is sequential and requires substantially more time than the simultaneous method. The microprobe has a mass resolution of about 1000, and a lateral imaging resolution of 2 μm. This instrument has provisions for viewing the sample during bombardment.

Recently, AEI has developed a primary ion-generating column, based on a design by Drummond and Long[42], which in conjunction with the AEI MS 702 constitutes a microprobe. While its technical specifications are equivalent to

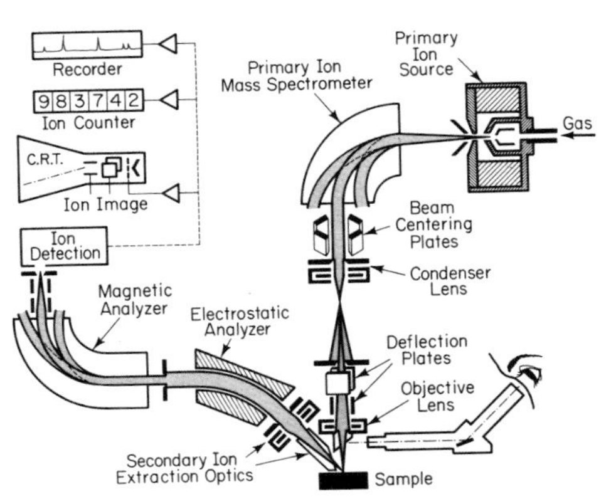


Fig. 14. Schematic diagram of the ARL scanning ion microprobe (after Evans[12]).

those of the ARL instrument, it has a substantially higher mass resolution: 3000 for electrical detection and 10 000 for photographic detection.

*3.6. Secondary ion mass spectrometry (SIMS)*

If good lateral resolution is not needed for surface analyses and in-depth profiling of thin films, the expensive ion probe microanalyzer just discussed can be replaced by the much simpler SIMS system. This usually consists of a primary ion gun of simple design, a sample manipulator, a secondary ion energy filter, and a quadrupole mass spectrometer, all mounted in an ultrahigh vacuum envelope. The major advantages of such a system are first, its relatively low cost ($25 000–$100 000) compared with either direct-imaging analyzers or ion microprobes (approximate price $300 000), and secondly, its ultrahigh vacuum capability, allowing primary ion beams of low current density to be used so that only a small fraction of a monolayer need be consumed in the study of a surface. The first successful instrument of this type was constructed by Benninghoven and Loebach[43]. In it, primary ions are produced in an axial ion gun, mass-analyzed magnetically, and focused into a 3 mm spot on the target. For analyses, primary beam densities in the range 1–10 nA $cm^{-2}$ are used at energies between 1 and 3 keV. This ion density removes only about 1 % of a monolayer during a "static" analysis lasting 1000 sec, with the system valved off the pumps. The secondary ions, positive and negative, emitted by the target are analyzed in a quadrupole mass spectrometer. Based on this design, a commercial system has been developed by Huber *et al.*[44] which utilizes ultrahigh vacuum techniques, including a Ti sublimation pump, to reduce residual gas pressures to below $10^{-10}$ torr. It uses a duoplasmatron source to produce primary ions focused directly, without mass analysis, into a 3 mm spot on the target. Thus current densities of about 10 μA $cm^{-2}$ are attained which permit fast sputtering rates for in-depth profiling.

Two other recent publications by Wittmaack *et al.*[45] and Schubert and Tracy[46] concerning secondary ion analyzers are based essentially on the Benninghoven design. Both groups found, however, that the simple ion gun–quadrupole combination produces a high continuous background which limits the dynamic detection range to at best four decades. For this reason, Wittmaack *et al.*[45] interposed a simple energy analyzer between target and quadrupole to prevent sputtered neutrals, light quanta and energetic scattered primary particles from entering the mass analyzer. By this means, a signal-to-noise ratio of $10^8$ was achieved in a favorable case. In their system, Schubert and Tracy[46] employed one-half of a cylindrical mirror analyzer to improve the signal-to-noise ratio to an estimated value of $10^6$ or better.

During 1974, a SIMS system was designed and constructed at RCA Laboratories by Magee[47] which will be ready for operation in mid-1975. As shown in Fig. 15, it consists of a commercially available Colutron high intensity ion source[48], followed by a Wien-type mass filter, a sample carrousel and manipulator, a spherical electrostatic analyzer for energy selection of the secondary ions, and a UTI 100C quadrupole mass analyzer. These components are housed in a differentially pumped ultrahigh vacuum system capable of reducing the pressure in the sample and mass analyzer regions to less than $10^{-8}$ torr during operation.

Among the strong points of SIMS are its high detection sensitivity, good depth resolution and positive mass identification. On the minus side are the severe matrix effects due to the presence of oxygen or other electronegative elements at the surface, which will affect secondary ion sensitivities by several orders of magnitude.

### *3.7. Auger electron spectroscopy*

Auger electron spectroscopy (AES), particularly as used in conjunction with ion sputtering, has been developed in recent years into one of the most effective tools for analyzing surfaces and thin films[9, 40]. Figure 16 is a diagram of the major components of a typical commercially available system[27]. These include a primary electron gun mounted axially in a cylindrical mirror system used to energy-analyze the Auger electrons, a carrousel sample holder and manipulator, one or two sputter ion guns used for depth profiling, an ultrahigh vacuum system to provide a contamination-free environment at a base pressure of about $10^{-10}$ torr, and associated electronic equipment. The latter includes a multiplex control which monitors six selected Auger peaks, permitting in-depth profiles to be obtained for six different elements. In the so-called thin film analyzer, the primary electron beam diameter ranges from about 100 μm to 25 μm, depending on primary current; a fixed sample area of primary beam size is analyzed, with an optimum detection sensitivity between 0.01 and 0.1%. The scanning Auger microprobe (SAM) recently developed by MacDonald[49] employs a smaller primary electron beam (15–4 μm diameter) which, together with electronic rastering circuitry, produces scanning electron microscope (SEM) pictures with a resolution of a few micrometers, as well as qualitative area scans and semi-quantitative line scans in the Auger mode. The last two capabilities provide concentration profiles within the plane of the surface for any selected element. The detection sensitivity for SAM is at the 0.1–1% concentration level.

### *3.8. X-ray photoelectron spectroscopy*

X-ray photoelectron spectroscopy (XPS), frequently called electron spectroscopy for chemical analysis (ESCA), requires complex instrumentation which

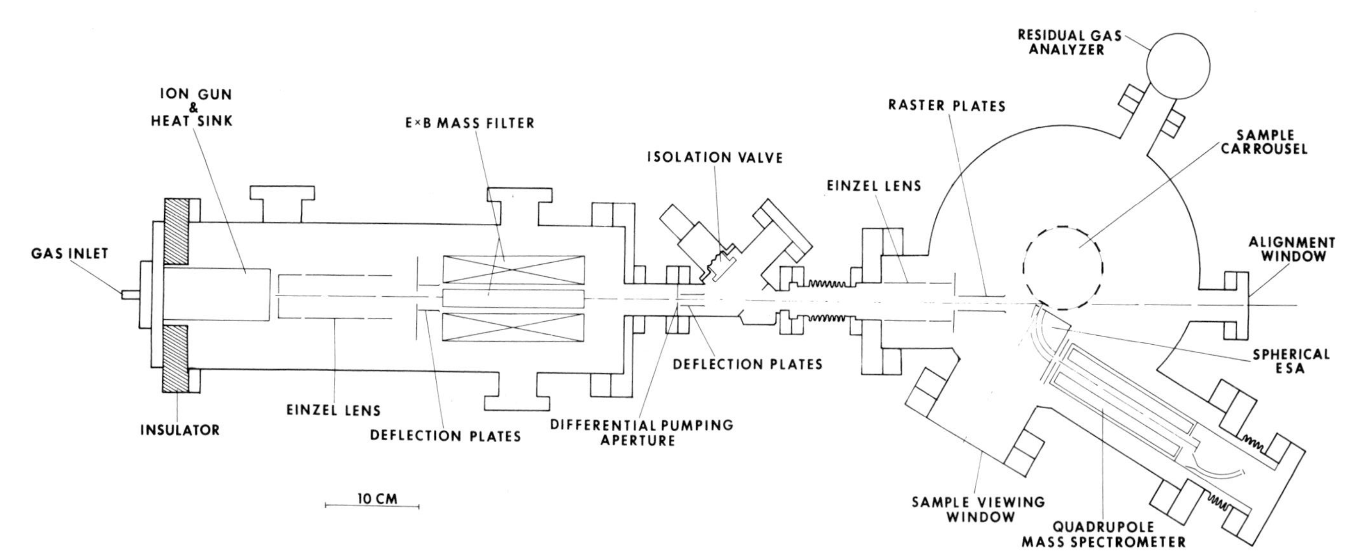


Fig. 15. Schematic diagram of a recent SIMS system designed and constructed at RCA Laboratories, ready for operation in 1975. It consists of the primary ion source followed by a Wien filter, a sample manipulator, a secondary ion electrostatic analyzer and quadrupole mass spectrometer, all housed in a differentially pumped vacuum system (after Magee[47]).

has only recently become generally available. Figure 17 shows a commercial instrument[50] of special design which optimizes the very small photoelectron currents. It consists of an X-ray source and monochromator, a sample manipulator, a dispersion-compensated electron lens and energy analyzer, and a large-area channel-plate detector and multichannel analyzer system which records the energy spectrum. This sophisticated and complex system is capable of identifying minor surface constituents at the 1% level and determining their molecular states, but the time required to carry out an analysis may amount to many hours.

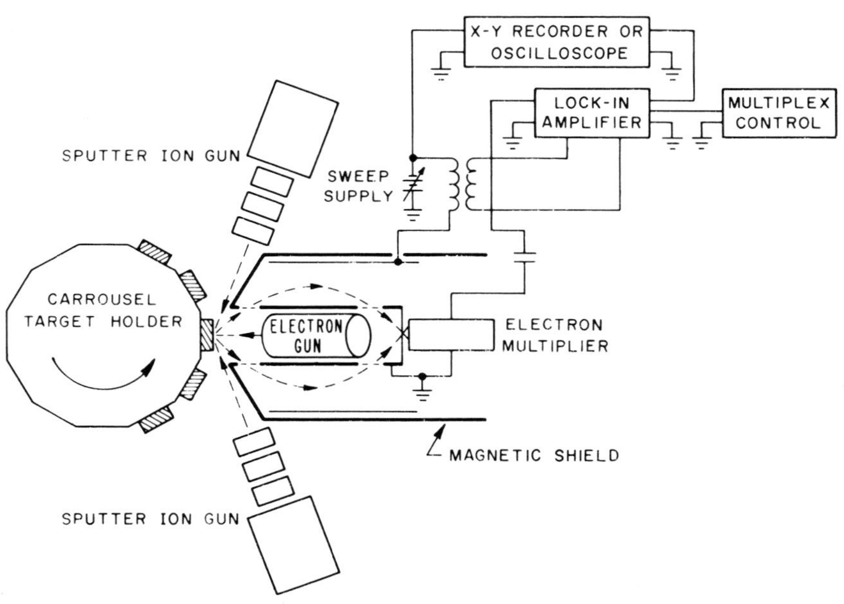


Fig. 16. Schematic diagram of an AES system, showing the cylindrical mirror analyzer, sample carrousel, ion sputter guns, electron detector and associated electronics (after PHI[27]).

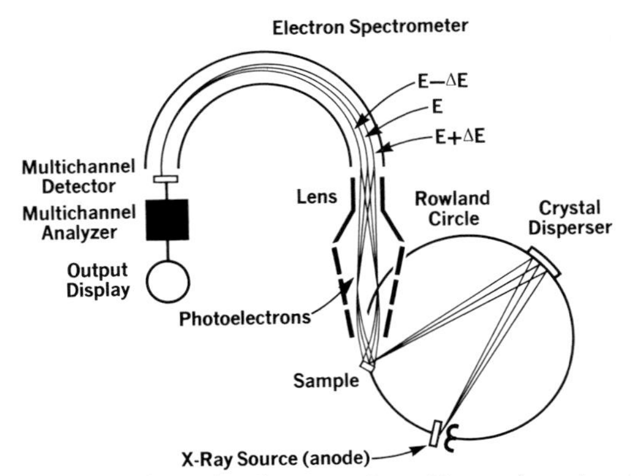


Fig. 17. Schematic diagram of an X-ray photoelectric spectrometer, showing the X-ray source and monochromator, the sample manipulator, the dispersion-compensated electron-optical system, the detector and the multichannel recording system (after Kelly and Tyler[50]).

## 4. RESULTS

In this section selected results will be presented to illustrate the different surface analytical methods described.

### *4.1. Mechanical top layer scan by SSMS*

In solid state technology it is of interest to check silicon wafers, at various stages of the processing cycle, for surface contaminants at the ppm level. This can be accomplished on the spark source mass spectrograph with the help of the automatic surface scanner described above. Tracks obtained[51] on a typical P-doped Si wafer with the help of a pointed gold counterelectrode are shown in Fig. 18. They consist of a series of craters approximately 5 μm deep which increase in width from 300 μm to 500 μm because the point of the counterelectrode broadens during the analysis. Multiple exposures were made to yield a detection limit of 0.1 ppma for most elements. The following impurities were found: in the range 0.1–1 ppma, Ag, Cr and Sb; in the range 1–10 ppma, B, C, Cl and Cu; and the dopant element P at 130 ppma. Similar analyses have been carried out on various samples selected from the entire Si processing cycle, and the conclusions reached were then used to modify the surface treatment at different stages.

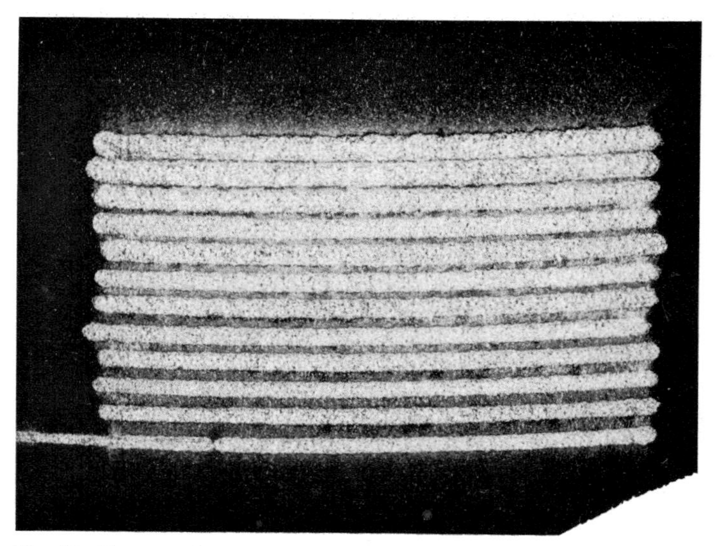

Fig. 18. Photograph of a SSMS top layer scan, showing tracks of sparked craters on a silicon sample. Track width varies from 300 μm to 500 μm, while the crater depth is about 5 μm.

### *4.2. Surface analysis of silicon by ISS*

The outermost atomic layers of ultra-pure, specially cleaned Si wafers of specific orientation have been examined[52] in some detail by ISS using both $He^+$ and $Ne^+$ primary ions. Figure 19 compares the spectrum of a (100) Si surface taken immediately after cleaning and minimum exposure to air (<10 min) with that obtained after ion-sputtering off about 30 Å of the surface layer. Surfaces had been prepared by cutting Si wafers in the desired plane and polishing them by standard techniques. The polished wafers were cleaned as follows: (1) boil in $4H_2O$:$H_2O_2$:$NH_4OH$, (2) $H_2O$ rinse, (3) boil in $4H_2O$:$H_2O_2$:HCl, (4) $H_2O$ rinse, (5) etch in concentrated HF, (6) steam oxidize at 1100 °C to an oxide thickness of about 3000 Å (this step oxidizes any structural surface damage due to polishing and contains within the oxide any surface impurities not removed by the previous etches), (7) repeat steps 1–5, (8) $H_2O$ rinse, being careful not to expose the Si surface to air until completion of the rinse. Ellipsometric measurements of oxide thickness were then made within 1–2 min and the wafers placed into the ISS vacuum chamber within 5–10 min. These surfaces were found to be optically damage-free, but the initial spectrum of Fig. 19 showed them to be largely covered with oxide, fluorine from the HF etch, and carbon (presumably an impurity in the etch). Of these impurities, only a trace of oxygen remained at the 30 Å level.

Spectra have been obtained for sputter-cleaned (111) Si and (100) Si surfaces which indicate that ISS yields concentrations that are quantitatively representative of surface atom density. From purely geometrical atom packing considerations, the computed (111):(100) density ratio is 1.15. Measured (111)Si:(100)Si concentration ratios were found to range from 1.09 to 1.19, agreeing with the expected value within experimental error.

*4.3. Determination of polar crystal orientation by ISS*

As already mentioned, the most unique characteristic of ISS is the fact that its detected signal derives predominantly from the topmost atomic layer of the solid. This capability is invaluable when the orientation of a polar crystal such as CdSe has to be determined. Figure 20 presents spectra obtained[53] for three different orientations of the polar crystal CdSe: the (0001) Cd face, the neutral

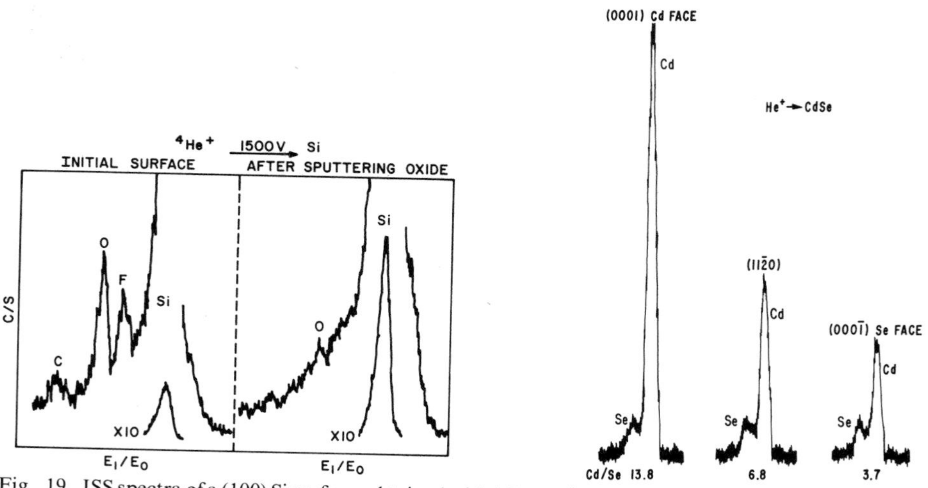


Fig. 19. ISS spectra of a (100) Si surface, obtained with 1500 eV $^4He^+$ ions. The left spectrum represents the initial surface, while the right one was taken after a 30 Å layer had been removed by sputtering.
Fig. 20. Effect of CdSe crystal orientation on ISS spectra.

$(11\bar{2}0)$ face and the $(000\bar{1})$ Se face. The data were taken with 1.5 keV $^4He^+$ ions. Even though the Cd intensities are substantially larger than the Se intensities for all three configurations, there are substantial differences between the two polar faces which make it possible to identify a given unknown crystal face. So far, no explanation is available for the unexpected disparity in elemental sensitivities observed for Cd and Se, but it was established that these sensitivities are strongly dependent on primary ion energy.

*4.4. Area and line scans by scanning Auger microanalysis*

The scanning Auger microanalyzer (SAM) is capable of producing area scans (maximum useful magnification 500×) and line scans (maximum useful magnification 1600×) showing the spatial distribution of any given element. Figure 21 is a composite picture[54] of a multiple nickel layer dipped in solder that had peeled off its silicon substrate. At 175× magnification, the photographs show SEM pictures of the secondary electron current and the spatial distributions of Ni, Sn and oxidized Si, respectively. When these area scans were examined in conjunction with separately obtained depth profiles, it was deduced that the Sn

SEM scan: emitted electrons.

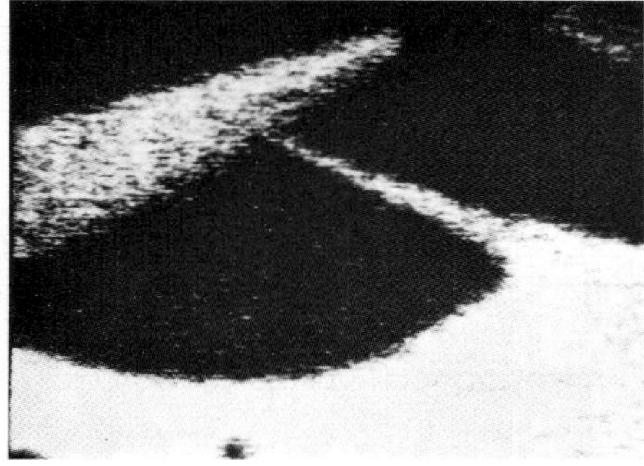

Auger scan: Ni 848 eV.

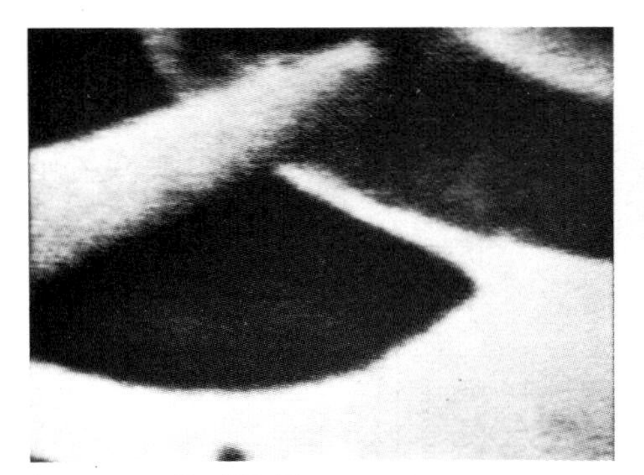

Auger scan: Sn 429 eV.

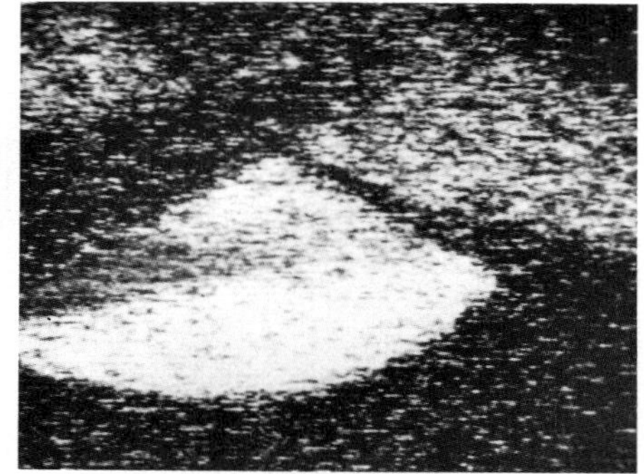

Auger scan: Si(ox) 1610 eV.

Fig. 21. SEM picture and Auger area scans (magnification 175×) obtained by SAM for Ni, Sn and Si of a solder-covered multiple nickel layer peeling off a silicon substrate. The pictures show clearly that the curled up portion of the layer consists of solder-diffused nickel, while the plane substrate is mainly oxidized silicon.

component of the solder had completely diffused through the multiple Ni layers, and that delamination had occurred at the nickel–silicon oxide interface in a patchy fashion.

Figure 22 is the tracing of a multiple line scan presenting lateral concentration profiles of four components for the device shown in cross section underneath the four profiles[54]. For the sake of clarity, the four zero levels are vertically displaced. The Cu profile taken at 918 eV indicates the presence of a metallic copper layer on the left, while the Si profile peaking at 1605 eV reveals an $SiO_2$ layer on the right. The 10 μm gap between these two layers is filled by oxidized aluminum, probably in the form of $Al_2O_3$, as evidenced by the fact that the Al line is a maximum at 1386 eV while it is zero at 1396 eV, the location of the metal peak. This interpretation is confirmed by the oxygen profile taken at 507 eV which rises from its zero level at the left, goes through a maximum in the gap, and remains at a steady high level in the $SiO_2$ region. From the shape of the Si profile, a lateral resolution of about 4 μm is calculated which is in good agreement with the minimum beam size claimed for SAM. These line scans were obtained at an electron beam current of 20 nA, and the measured energy values quoted are accurate to better than ±2 eV.

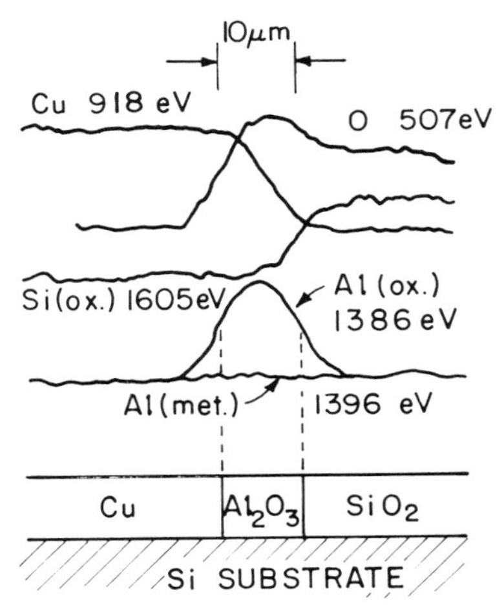


Fig. 22. Auger line scans of metallic Cu, oxidized Al and Si, and O obtained by SAM for the multilayer device shown in diagrammatic form underneath the profiles (magnification 800×). A lateral resolution of about 4 μm is deduced from the Si profile.

*4.5. In-depth concentration profiling by EPM*

As discussed earlier, the electron probe microanalyzer (EPM) has a typical information depth of 0.3–1 μm; thus its depth resolution is not adequate for many thin film problems. However, when this method is applied to angle-lapped specimens, approximate depth concentration profiles can be obtained[55] even for complex multilayer structures, as illustrated in Fig. 23. A 1° angle-lap is equivalent to a lateral amplification factor of 57; thus a layer of thickness 1 μm will appear to be a band of width 57 μm. The Al profile of Fig. 23, when compared with the layer boundaries, demonstrates the limitations of this method—because of the considerable penetration depth of the 15 keV primary electron beam used, the changes in Al concentrations at the various interfaces appear to be laterally displaced and gradual instead of sharp. Nevertheless, the method is useful for the semi-quantitative exploration of complex structures.

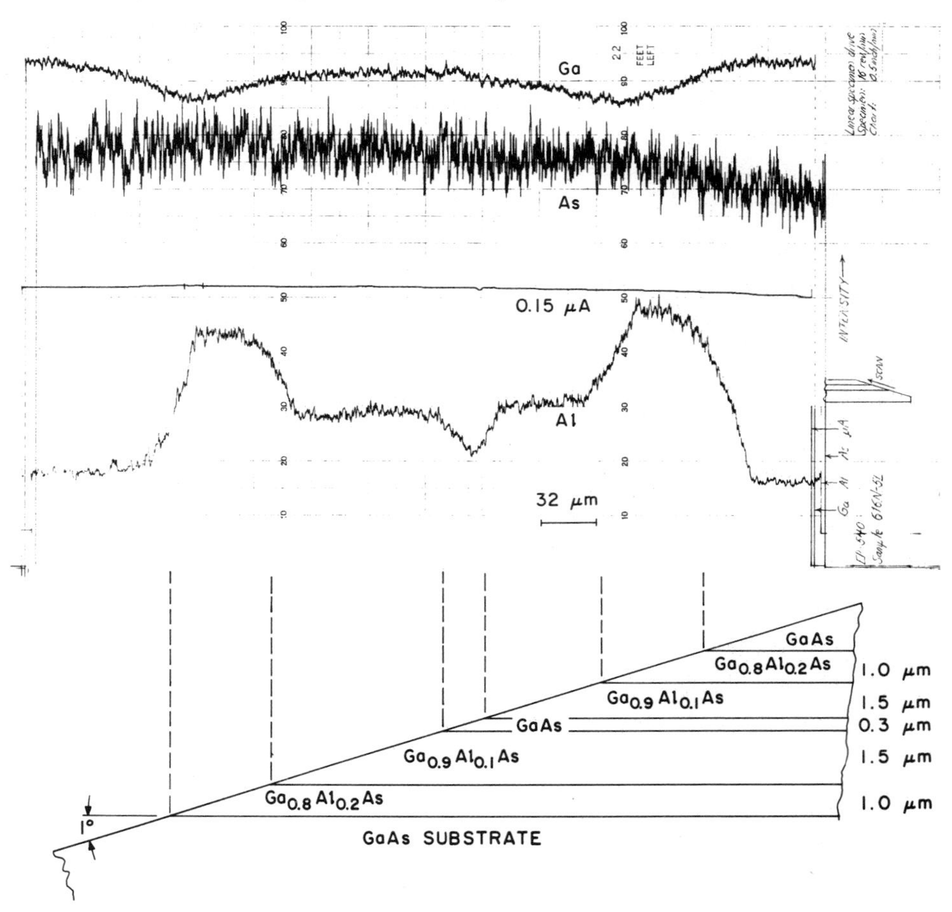


Fig. 23. Electron probe microanalysis of an angle-lapped specimen of a complex multilayer structure. For a 1° angle-lap the lateral amplification is 57, permitting layers of micrometer thickness to be explored semi-quantitatively.

### *4.6. Depth profile of a boron-in-silicon implant by IPM*

Ion probe microanalysis (IPM) is particularly effective as a method for the determination of ion concentrations implanted in a solid. As an example we shall consider the case of $^{11}B^+$ ions of 150 keV energy implanted into a Si wafer (5–10 Ω cm, n-type). A total dose of $2.176 \times 10^{15}$ ions cm$^{-2}$ was implanted, corresponding to a maximum B concentration of $1 \times 10^{20}$ atoms cm$^{-3}$ with a computed range $R_p = 0.4704$ μm and a straggling parameter $\Delta R_p = 0.0868$ μm. These values apply to the unannealed case and have been computed[56] using realistic values of electronic stopping powers rather than LSS theory. Measurements were made[57] on a Cameca direct-imaging mass analyzer using a primary beam of 5.5 keV $O_2^+$ ions, 150 μm in diameter, which was rastered over a 1 $mm^2$ area in order to produce a flat-bottomed crater. The detection of secondary ions was limited to those emitted by the central region 200 μm in diameter. Figure 24 shows the experimentally determined **B** concentration profile for the unannealed case (crosses), the maximum being adjusted to the computed value of $1 \times 10^{20}$ atoms $cm^{-3}$. The experimental range $R_p = 0.467$ μm and straggling parameter $\Delta R_p = 0.096$ μm are in good agreement with the computed values quoted above, inspiring confidence in the method. The depth scale was obtained by determining the crater profile on a Talysurf 4 profilometer. After annealing the sample for 15 min at 1000 °C (circles), the maximum concentration had dropped substantially. The range $R_p$ remained essentially the same, while the straggling parameter $\Delta R_p$ increased significantly.

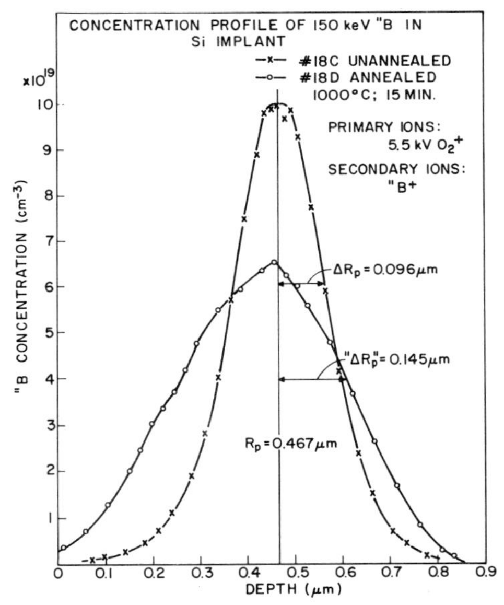


Fig. 24. Concentration profile of a 150 keV $^{11}B$ in Si implant. Total dose $2.176 \times 10^{15}$ atoms $cm^{-2}$; maximum unannealed concentration $1 \times 10^{20}$ atoms $cm^{-3}$. Taken on a Cameca IPM with 5.5 kV $O_2^+$ ions.

### 4.7. *In-depth concentration profiles of a Cu/Mo/Si sample by AES, ISS and IPM*

To compare the capabilities and limitations of the three major surface analytical techniques, a special multilayered structure was analyzed by Auger electron spectroscopy, by ion scattering spectrometry and by ion probe microanalysis. The sample consisted of a Cu layer of thickness 965 ± 50 Å, a Mo layer of thickness 1530 ± 50 Å and a Si substrate. Layer thicknesses had been independently determined by interferometry and by Talysurf 4 profilometry, yielding results that agreed to within the limits quoted. Figure 25 presents the most significant in-depth concentration profiles obtained by each of the methods listed above. The percentage figures indicated in Figs. 25(a) and 25(b) represent depth resolution values at the interfaces, and show that in this instance AES yielded better results than raster-gated ISS. These computed values are based on the definitions presented in Section 2.2 and take into account the substantial differences in sputtering rates which are deduced from the measured times and layer thicknesses. Still, the depth resolution obtained by AES on this particular sample (Fig. 25(a)) is substantially worse than that obtained on another similar sample which yielded a value of about 12%. Thus we must conclude that the two interfaces in this sample were not well defined, probably due to interdiffusion of the constituents.

The in-depth concentration profile obtained on the Cameca ion probe (Fig. 25(c)) bears little resemblance to the two profiles obtained by AES and ISS. The Cu concentration appears to go up dramatically near the Cu–Mo interface, and the Si signal reaches its maximum level well ahead of the Mo–Si interface. These unexpected changes are artifacts caused by the presence of about 15 at.% oxygen in the Mo layer, as determined by the AES analysis. These artifacts are explained by the strong dependence of secondary ion emission on the presence of oxygen in the lattice (see Section 2.1.3), even though in this case the ion probe profile was obtained with primary $O_2^+$ ions. These results clearly indicate the practical difficulties encountered when secondary ion mass analysis is employed: to attain optimum detection sensitivity it is desirable to employ primary $O_2^+$ ions; on the other hand, this makes it impossible to determine an oxygen concentration profile for the sample. Thus a profile measured by this method for a given constituent may either reflect its actual concentration, or be an indication of varying concentrations of oxygen being present in the lattice.

### *4.8. Intercomparison and evaluation of surface and thin film methods*

The capabilities and limitations of the various methods discussed in this paper are compared in terms of the more important practical parameters in Table III. Such a tabulated comparison, while concise, if of necessity incomplete and difficult to achieve because methods are forced into a common mold that is only partially applicable. To complement Table III, major capabilities and limitations are listed below in capsule form for each method.

X-ray fluorescence spectrometry (XRFS), electron probe microanalysis (EPM) and spark source mass spectrography (SSMS) are three methods with sampling depths in the micrometer range. XRFS is non-destructive, quantitative, fast, fairly sensitive, and readily applied to insulators; it has no lateral resolution and cannot detect elements below fluorine. EPM is usually non-destructive, fast, quantitative, and yields area and line scans with excellent lateral resolution (about 1 μm); its detection sensitivity for elements below fluorine is very limited. SSMS is an excellent survey method of high sensitivity (often 0.01 ppma or better) which positively identifies all elements via their isotopic abundance ratios; it can be made quantitative through electrical read-out, but has only limited lateral resolution.

Ion scattering spectrometry (ISS) determines the composition of the outermost one or two surface layers, which may differ substantially from the under-

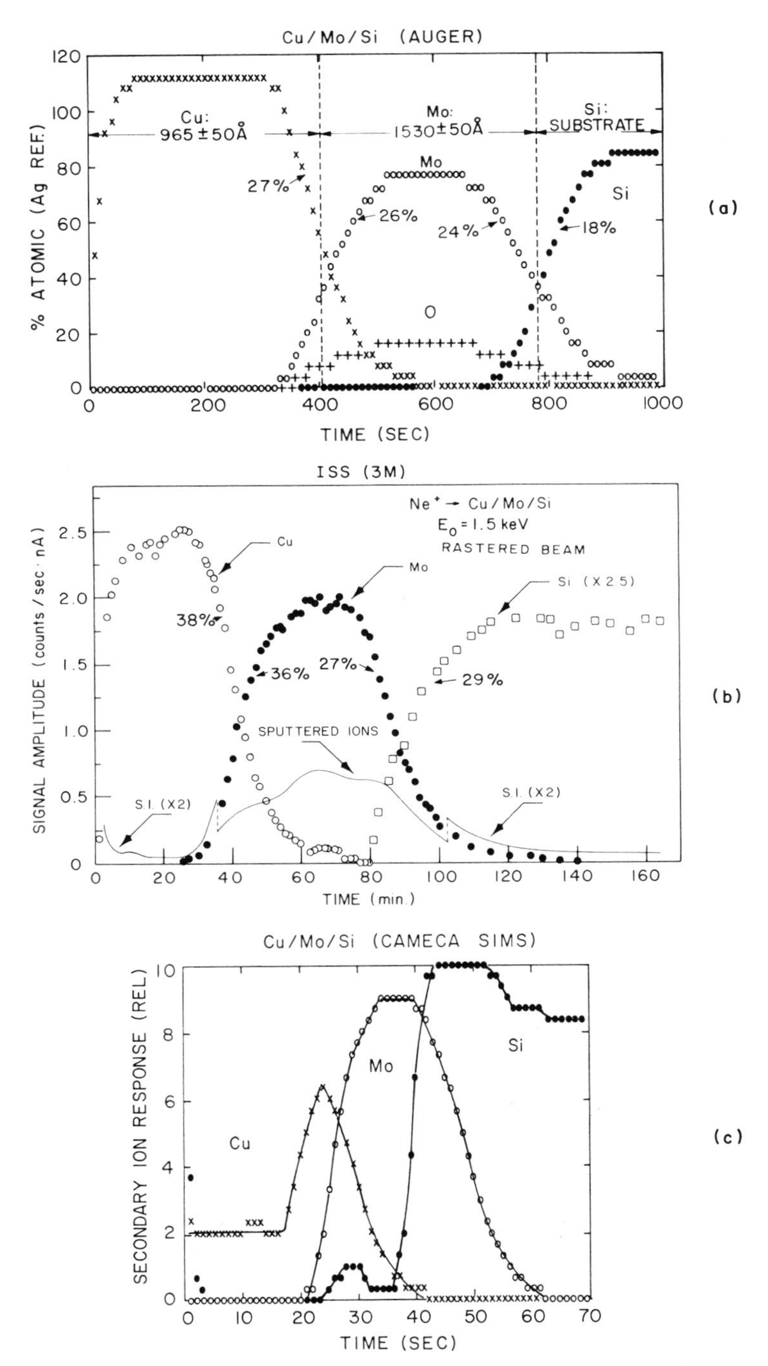


Fig. 25. In-depth concentration profiles of a Cu/Mo/Si sample obtained by AES, ISS and IPM. Percentage figures at the interfaces indicate the measured depth resolution values.

TABLE III

SURFACE AND THIN FILM METHODS: CAPABILITIES AND LIMITATIONS

| METHOD | DETECTION SENSITIVITY | | EFFECTS | | CHARGE – UP & FIELD PROB-LEMS | BEAM INDUCED CHEM'L CHANGES | DEPTH RES. RASTER/GATING | | LATERAL RESOL'N | ELEM'L IDENT'N | TYP-ICAL ANAL'L TIME | CAPABILITIES LIMITATIONS |
|---|---|---|---|---|---|---|---|---|---|---|---|---|
| | OPTIMUM (ppma) | RANGE FACTOR | MATRIX | GEOME-TRICAL | | | WITH-OUT | WITH | | | | |
| X-RAY FLUORESCENCE SPECTROMETRY | 1-100 | | | | NO | | | | NONE | GOOD | 15 min | C: NON-DESTRUCTIVE; QUANTITATIVE; FAST; L: Z>9 |
| ELECTRON-PROBE MICROANALYSIS | 100-1000 | 10 | SOME | YES | YES | YES | | | EXC. | GOOD | 1 h | C: "NON-DESTRUCTIVE;" QUANTITATIVE; AREA AND LINE SCANS L: Z≥4 |
| SOLIDS MASS SPECTROGRAPHY | 0.01-10 | 10 | | | YES | | | | FAIR | EXC. | 1 h | C: SENSITIVE SURVEY METHOD |
| ION SCATTERING SPECTROMETRY | 0.1-1% | 10 | | YES | NO | YES | POOR | FAIR | POOR | FAIR | 3 h | C: TRUE "SURFACE" ANALYSIS; INSULATORS; DEPTH PROFILE |
| SECONDARY ION MASS SPECTROMETRY | 0.1-100 | $10^4$ | SEVERE | | YES | YES | POOR | GOOD | FAIR | GOOD | 30 min | C: DEPTH PROFILE; L: MATRIX EFFECTS |
| ION-PROBE MICROANALYSIS | 0.1-100 | $10^4$ | SEVERE | YES | YES | YES | POOR | GOOD | GOOD | GOOD | 30 min | C: AREA AND LINE SCANS; DEPTH PROFILE; L: MATRIX EFFECTS |
| AUGER ELECTRON SPECTROMETRY | 0.01-0.1% | 20 | | | YES | YES | GOOD | | FAIR | GOOD | 30 min | C: DEPTH PROFILE MULTIPLEXING OF SIX ELEMENTS |
| "SAM": SCANNING AUGER MICROANALYSIS | 0.1-1% | 20 | | | YES | YES | GOOD | | GOOD | GOOD | 1 h | C: AREA AND LINE SCANS; DEPTH PROFILES OF SIX ELEMENTS |
| "XPS": X-RAY PHOTOELECTRON SPECTROSCOPY ("ESCA") | 0.1-1% | | | | NO | | | | NONE | FAIR | 3h | C: MOLECULAR INFORMATION: VALENCE STATES L: SLOW METHOD |

lying layers because of preferential sputtering. Major and minor constituents are identified with modest mass and lateral resolution, while its in-depth resolution has been improved with the help of raster-gating of the ion beam. Sample consumption and sputtering rate are both low, making this an attractive method for thin films. It can analyze insulators and establish the orientation of polar crystals.

Secondary ion microanalyzers (IPM), both direct-imaging and scanning, have excellent mass, lateral and in-depth resolution capabilities and high sensitivity, at the expense of high sample consumption. The analytical information comes from a few layers below the sample surface and represents bulk concentrations at that level. Their major limitations are the strong effect of surface and bulk oxygen on ion intensities (" matrix effect "), and the wide range of secondary positive ion yields which differ from element to element by as much as four orders of magnitude. Thus in-depth profiles are frequently hard to evaluate quantitatively.

Secondary ion mass analyzers (SIMS) have good mass and in-depth resolution, but poor lateral resolution. Their analytical information comes from a modest depth and their sample consumption is potentially low, sometimes amounting to only 1% of a monolayer. Matrix effect and secondary ion yield problems apply equally to this method.

Auger electron spectrometry (AES) and scanning Auger microanalysis (SAM), when used in conjunction with ion sputtering, are capable of producing quantitative in-depth concentration profiles for up to six constituents via multiplexing. These two methods have only limited detection sensitivity (0.01–0.1 at.%), but are fast and essentially free of matrix effects. In addition, SAM provides area and line scans with an optimum resolution of about 4 μm. In certain samples, especially compounds with high vapor pressure, the electron beam may induce chemical changes.

X-ray photoelectron spectroscopy (XPS or ESCA) is the only method capable of providing, via electronic bonding energies, information concerning the chemical structure of atoms located within the first few layers of a solid surface. The method can be applied to insulators but it is slow and has no lateral resolution. Its detection sensitivity is limited to about 1%.

From the foregoing comparisons and evaluations the following conclusion may be drawn. There does not exist at present any single method for the analysis of surfaces and thin films which is applicable to all types of materials, including insulators, and which exhibits high detection sensitivity and three-dimensional resolution, adequate reproducibility, low sample consumption, and complete elemental coverage. To solve a given problem it is therefore necessary to establish first the major requirements and constraints, and then to select the best-suited method or methods.

## ACKNOWLEDGMENTS

The author acknowledges with pleasure many helpful discussions with a number of his colleagues at RCA Laboratories, including E. P. Bertin, E. M. Botnick, D. G. Fisher, W. L. Harrington, C. W. Magee and R. J. Paff. For making available their original figures, he is especially grateful to D. R. Beaman, Dow Chemical, Midland; E. P. Bertin, RCA Laboratories; C. A. Evans, Jr., University of Illinois; D. G. Fisher, RCA Laboratories; M. A. Kelly, Hewlett-Packard; N. C. MacDonald, Physical Electronics Industries; and C. W. Magee, RCA Laboratories.

## REFERENCES

1 P. F. Kane and G. B. Larrabee (eds.), *Characterization of Solid Surfaces*, Plenum Press, New York, 1974.
2 D. R. Beaman and J. A. Isasi, *Electron Microbeam Analysis*, ASTM Special Technical Publication 506, Philadelphia, Pa., 1972.
3 C. C. Chang, Analytical Auger spectroscopy. In P. F. Kane and G. B. Larrabee (eds.), *Characterization of Solid Surfaces*, Plenum Press, New York, 1974, Chap. 20.
4 R. E. Honig, Analysis of surfaces and thin films by mass spectrometry. In A. R. West (ed.), *Advances in Mass Spectrometry*, Vol. 6, Elsevier Appl. Sci. Publ., Barking, England, 1974, pp. 337–362.
5 G. Carter and J. Colligon, *Ion Bombardment of Solids*, McGraw-Hill, London, 1968.
6 G. K. Wehner, *Sputtering Yield Data in the 100–600 eV Energy Range*, General Mills Report No. 2309, July 1962.
7 M. L. Tarng and G. K. Wehner, *J. Appl. Phys.*, *42* (1971) 2449.
8 S. D. Dahlgren and E. D. McClanahan, *J. Appl. Phys.*, *43* (1972) 1514.
9 P. W. Palmberg, *Anal. Chem.*, *45* (1973) 549A.
10 G. Slodzian and J. F. Hennequin, *C. R. Acad. Sci.*, *Ser. B*, *263* (1966) 1246.
11 R. Castaing and J. F. Hennequin, in A. Quayle (ed.), *Advances in Mass Spectrometry*, Vol. V, Institute of Petroleum, London, 1972, pp. 419–424.

12 C. A. Evans, Jr., *Anal. Chem.*, *44* (Nov. 1972) 67A.
13 C. A. Evans, Jr., *Proc. 8th Natl. Conf. of the Electron Probe Analysis Soc. Am., New Orleans, La., 1973.*
14 R. Castaing and G. Slodzian, *J. Microsc. (Paris)*, *1* (1962) 395.
15 A. Benninghoven, *Z. Naturforsch., Teil A*, *22* (1967) 841.
16 C. A. Andersen, *Int. J. Mass Spectrom. Ion Phys.*, *2* (1969) 61.
17 C. A. Andersen, *Int. J. Mass Spectrom. Ion Phys.*, *3* (1970) 413.
18 C. A. Andersen, *Anal. Chem.*, *45* (1973) 1421.
19 J. A. McHugh, Secondary ion mass spectrometry. In S. P. Wolsky and A. W. Czanderna (eds.), *Methods of Surface Analysis*, Elsevier, Amsterdam, 1975.
20 R. K. Lewis, J. M. Morabito and J. C. C. Tsai, *Appl. Phys. Lett.*, *23* (1973) 260.
21 A. Benninghoven and A. Mueller, *Phys. Lett. A*, *40* (1972) 169.
22 A. Benninghoven, C. Plog and N. Treitz, *Int. J. Mass Spectrom. Ion Phys.*, *13* (1974) 415.
23 V. E. Krohn, Jr., *J. Appl. Phys.*, *33* (1962) 3523.
24 A. Socha, *Surf. Sci.*, *25* (1971) 147.
25 J. M. Morabito and R. K. Lewis, *Anal. Chem.*, *45* (1973) 869.
26 R. F. Goff and D. P. Smith, *J. Vac. Sci. Technol.*, *7* (1970) 1.
27 Physical Electronics Industries, Technical Information, 1975.
28 E. P. Bertin, X-ray secondary emission (fluorescence) spectrometry. In *Principles and Practice of X-Ray Spectrometric Analysis*, Plenum Press, New York, 1970, Chap. 3.
29 J. V. Gilfrich, X-ray fluorescence analysis. In P. F. Kane and G. B. Larrabee (eds.), *Characterization of Solid Surfaces*, Plenum Press, New York, 1974, Chap. 12.
30 E. P. Bertin, The electron-probe microanalyzer. In *Principles and Practice of X-Ray Spectrometric Analysis*, Plenum Press, New York, 1970, Chap. 21.
31 W. Reuter, *Surf. Sci.*, *25* (1971) 80.
32 C. A. Andersen (ed.), *Microprobe Analysis*, Wiley, New York, 1973, Chaps. 1–11.
33 G. A. Hutchins, Electron probe microanalysis. In P. F. Kane and G. B. Larrabee (eds.), *Characterization of Solid Surfaces*, Plenum Press, New York, 1974, Chap. 18.
34 J. M. McCrea, Mass spectrometry. In P. F. Kane and G. B. Larrabee (eds.), *Characterization of Solid Surfaces*, Plenum Press, New York, 1974, Chap. 21.
35 C. W. Magee and W. W. Harrison, *Anal. Chem.*, *45* (1973) 220.
36 C. W. Magee, Personal Communication, 1974.
37 R. F. Goff, *J. Vac. Sci. Technol.*, *10* (1973) 355.
38 T. W. Rusch, J. T. McKinney and J. A. Leys, submitted to *J. Vac. Sci. Technol.*
39 H. Liebl, *J. Appl. Phys.*, *38* (1967) 5277.
40 J. M. Morabito, *Thin Solid Films*, *19* (1973) 21.
41 C. A. Andersen and J. R. Hinthorne, *Science*, *175* (1972) 853.
42 I. W. Drummond and J. V. P. Long, *Nature (London)*, *215* (1967) 950.
43 A. Benninghoven and E. Loebach, *Rev. Sci. Instrum.*, *42* (1971) 49.
44 W. K. Huber, H. Selhofer and A. Benninghoven, *J. Vac. Sci. Technol.*, *9* (1972) 482.
45 K. Wittmaack, J. Maul and F. Schulz, *Int. J. Mass Spectrom. Ion Phys.*, *11* (1973) 23.
46 R. Schubert and J. C. Tracy, *Rev. Sci. Instrum.*, *44* (1973) 487.
47 C. W. Magee, Personal Communication, 1975.
48 M. Menzinger and L. Wahlin, *Rev. Sci. Instrum.*, *40* (1969) 102.
49 N. C. MacDonald, The third dimension in scanning electron microscopy: Scanning Auger microscopy. In D. R. Beaman and B. Siegel (eds.), *Electron Microscopy: Physical Aspects*, Wiley, New York, 1975.
50 M. A. Kelly and C. E. Tyler, *Hewlett-Packard Journal* (July 1973) 2.
51 E. M. Botnick, Personal communication, 1974.
52 W. L. Harrington and R. E. Honig, Paper presented at *22nd Annu. Conf. on Mass Spectrometry and Allied Topics, Philadelphia, Pa., May 1974.*
53 R. E. Honig and W. L. Harrington, *Thin Solid Films*, *19* (1973) 43.
54 D. G. Fisher, Personal communication, 1975.
55 E. P. Bertin, Personal communication, 1974.
56 C. P. Wu and E. C. Douglas, Personal communication, 1974.
57 W. L. Harrington and C. W. Magee, Personal communication, 1975.

*Thin Solid Films*, *31* (1976) 123–138



# ELECTRICAL AND GALVANOMAGNETIC MEASUREMENTS ON THIN FILMS AND EPILAYERS

H. H. WIEDER
*Naval Electronics Laboratory Center, San Diego, Calif. 92152 (U.S.A.)*
(Received May 12, 1975; accepted June 17, 1975)

Measurements of resistivity and Hall coefficient are necessary for the determination of carrier concentration and mobility in thin films and epilayers. The measurement techniques are reviewed, with particular emphasis on the role of perturbations and errors introduced into the measurements by the finite size of electrodes, by spatially ordered or randomly dispersed conductive or non-conductive inclusions and by inhomogeneities arising during layer growth, impurity diffusion or ion implantation.

## 1. ROLE OF SPECIMEN CONTOUR

The resistivity $\rho$ and the Hall coefficient $R_h$ of arbitrarily shaped laminae can be measured by the method developed by van der Pauw[1, 2]. The method depends on the conformal representation of the specimen on an infinite half-plane of thickness $d$; four peripheral line electrodes are oriented along $d$. Points A and B are the projections on the plane of two of the electrodes, respectively the source and sink of an applied current $i$. The current density with respect to each electrode at a radial distance $r$ is $J = i/2\pi rd$; the current streamlines are radial and the equipotentials are concentric circles.

If the resistivity is a scalar quantity then in $H = 0$ the potential difference measured between the other two line electrodes whose projections are designated by C and D, together with the current $i$, determine the resistance $R_{AB,CD}$. Commutating the current and potential electrodes and repeating the measurements leads to the resistance $R_{BC,DA}$. The resistivity is then calculated from the solution of Laplace's equation for the potential distribution in the plane subject to the appropriate boundary conditions:

$$f(\rho) = \exp(-\pi d R_{AB,CD}/\rho) + \exp(-\pi d R_{BC,DA}/\rho) - 1 = 0 \qquad (1)$$

van der Pauw has shown that any arbitrarily shaped lamina, which can be mapped by conformal transformation onto an infinite half-plane, yields analytically the same eqn. (1) provided that:

(a) $\nabla \cdot \boldsymbol{J} = 0 \qquad \nabla \times \boldsymbol{J} = 0$

(b) the lamina is simply-connected;

(c) it is homogeneous and isotropic;
(d) it is uniform in thickness;
(e) its four "ohmic" line electrodes are on its periphery;
(f) the line-electrode projections on the surface of the lamina are point contacts.

If these electrodes are disposed along orthogonal diameters of a circular specimen, as shown in Fig. 1(a), or at the corners of a square, then $R_{AB,CD} = R_{BC,DA} = R$; the resistivity can then be determined from a single measurement, that of the potential difference between the potential probes, and the known input current $i$. Equation (1) yields the resistivity

$$\rho = \pi d R / \ln 2 \tag{2}$$

The same specimen can be used for the evaluation of the Hall coefficient by applying a magnetic field $H$ transverse to the lamina, while the current $i$ is applied to A and C and the potential difference is measured between B and D. If B and D are on an equipotential line, then the voltage $v_0$ between them is zero for $H = 0$. For $H \neq 0$ the equipotentials are rotated as shown in Fig. 1(b), and the Hall voltage between B and D is

$$v_h = R_h H i / d \tag{3}$$

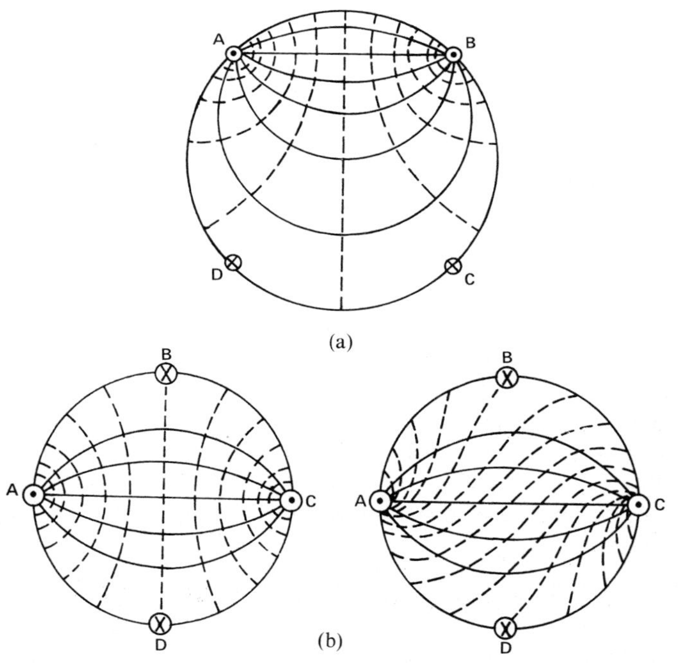


Fig. 1. Distribution of current streamlines (continuous lines) and equipotentials (dotted lines): (a) electrode configuration for resistivity measurements, $H = 0$; (b) electrode configuration for Hall measurements; on the left $H = 0$, on the right $H$ is normal to the plane of the page. Current electrodes are identified by dots, potential probes by crosses.

However, B and D are not, as a rule, on an equipotential for $H = 0$ and a "misalignment voltage" appears between them. Therefore, in $H \neq 0$ the measured potential difference is $v_0 = v_r \pm v_h$. The effective Hall voltage can be extracted from two Hall measurements using the same input current but equal and oppositely oriented magnetic fields $H_+$ and $H_-$; the Hall voltage is then half the algebraic sum of the corresponding $v_{0+}$ and $v_{0-}$. Thermoelectric potentials can also introduce errors; they can be reduced by performing two measurements in a constant magnetic field with opposite polarity, equal amplitude currents. The corresponding output potentials are $v_0(+)$ and $v_0(-)$. Then, neglecting thermo-galvanomagnetic effects considered to be of second order, the effective Hall voltage determined from all four measurements is

$$v_h = \tfrac{1}{4}\{v_{0+}(+) - v_{0-}(+) + v_{0-}(-) - v_{0+}(-)\} \tag{4}$$

Small area electrodes are usually deposited onto film or epitaxial layer surfaces rather than the line electrodes required for van der Pauw measurements. If the diameter of the electrodes is much smaller than the specimen diameter then a point-contact approximation may be valid. A symmetrical four-point probe array can then be treated analytically by the method of "Corbino images" developed by Buehler and Pearson[3] and by Buehler[4]. The current streamlines and equipotentials near a point source or sink are similar to those of a Corbino disc if the point source is on an infinite plane; the equipotentials are invariant in $H$ and the current streamlines are rotated as a function of field, producing a large magnetic field-dependent "geometrical origin" magnetoresistance. For a Corbino source at the edge of an infinite half-plane the current streamlines are invariant in $H$ and the equipotentials are rotated in accordance with eqn. (3). The Corbino source analysis of a symmetrical four-point probe peripheral array supported by experimental data[3, 4] shows that $\rho$ and $R_h$ can be calculated from eqns. (2) and (3), just as for the van der Pauw measurements.

The resistivity and Hall coefficients of symmetrical thin films or layer specimens can be determined from such four-point probe measurements with high accuracy provided that the boundary conditions (points (a)–(f) above) are met. Violations of these boundary conditions introduce errors in $\rho$ and $R_h$ which are then reflected in the derived fundamental charge carrier transport parameters.

## 2. ERRORS INTRODUCED BY ELECTRODE SIZE, GEOMETRY AND POSITION

The errors introduced by finite electrodes or by the displacement of an electrode from the specimen periphery were evaluated by van der Pauw[1, 2] for circular specimens; he indicated that these errors can be minimized by choosing an appropriate specimen contour such as the "clover-leaf" shape. However, such a geometrical structure requires complex preparation procedures; a clover-leaf configuration applied to films or epitaxial layers requires complex photolithographic processing which may alter their surfaces. Such a structure also occupies too much of the useful area of a specimen, removing it from consideration for other applications. For this reason, simple circular or square structures with simple electrode configurations were considered to be desirable. Chwang *et al.*[5] have investigated both analytically and experimentally the errors introduced in $\rho$ and $R_h$ by triangular or square peripheral surface electrodes applied at the corners of square specimens, as shown in Fig. 2. Experimental measurements were made in an appropriate electrolytic tank equipped with suitable electrodes, in order to determine the analog potential distribution; the measurements were compared with theoretical calculations solving Laplace's equation by the method of finite differences applied to a resistive analog circuit with appropriate boundary conditions. They calculated the sheet resistivity $\rho d$ as a function of the electrode dimension $\delta$ and the specimen edge length $l$ shown in Fig. 2. The experimental data were found to be in good agreement with theory; since $v_0$ measured between finite potential probes is always smaller than that between idealized point-contact electrodes, the measured resistivity $\rho_m$ is always smaller than the actual resistivity $\rho$. For triangular electrodes, the resistivity correction $\rho/\rho_m$ appears to increase monotonically with the ratio $\delta/l$. For conditions usually encountered in practice $l$ may be greater by a factor of 5 to 10 in comparison with $\delta$ and the errors introduced by the contact size are quite small. For example, for $\delta/l \simeq 0.125$, $\rho/\rho_m = 1.0025$ for triangular contacts; for the same $\delta/l$ and square contacts $\rho/\rho_m \simeq 1.004$.

The errors introduced by finite electrodes in Hall measurements can be attributed to two causes: (a) the electrodes short-circuit the Hall field and therefore reduce the measured Hall voltage in comparison with that obtained with point-contact electrodes; (b) finite electrodes short-circuit the current streamlines near and around Hall electrodes and can therefore introduce non-linearities in the magnetic field dependence of $v_h$ in eqn. (3). The Hall voltage correction factor relating the effective to the measured Hall voltage $v_h/v_{hm}$ is greater than that of the resistivity, and roughly proportional to $(\delta/l)^2$. For $\delta/l \simeq 0.125$, $v_h/v_{hm} \simeq 1.10$ for triangular electrodes. Defining the Hall angle $\theta = R_h H/\rho$, the correction factor has a relatively weak dependence on the Hall angle for $0.1 \leqq \tan\theta \leqq 0.5$; for $\delta/l \simeq 0.125$, $v_h/v_{hm} = 1.10 \pm 0.1$.

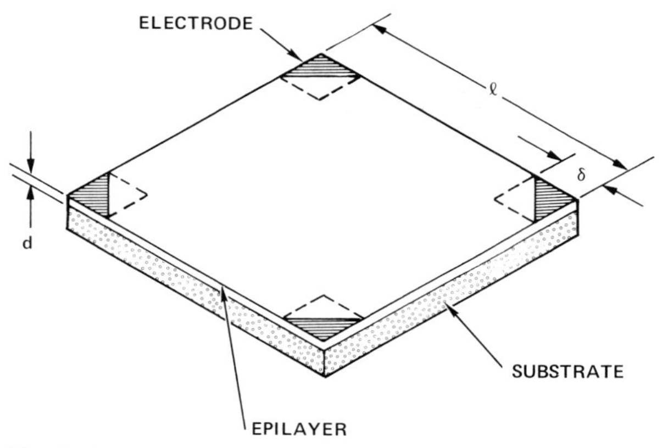


Fig. 2. Square contour epilayer of edge length $l$ and thickness $d$ on its semi-insulating substrate; triangular (square shown by broken lines) electrodes have an edge length $\delta$.

The correction factors of finite size circular electrodes of diameter $\delta$, symmetrically disposed on the periphery of circular germanium specimens of diameter $l$, were investigated experimentally by van Daal[6]. He found that $\rho/\rho_m$ is proportional to $\delta/l$. For $\delta/l \simeq 0.125$ and a specimen thickness $\alpha = 0.245$ cm he found that $\rho/\rho_m \simeq 1.03$ and $v_h/v_{hm} \simeq 0.20$. These values are considerably larger than those calculated and measured on square specimens with triangular electrodes. Furthermore, they suggest a possible dependence of the resistivity and Hall correction factors on specimen thickness.

## 3. SYMMETRY AND ASYMMETRY OF ELECTRODES

If four peripheral surface electrodes are not equispaced on a circular specimen then, as a rule, eqns. (2) and (3) are not applicable. Buehler and Pearson[3] and Buehler[4] have shown that the resistivity of a lamina with an asymmetrical electrode disposition, shown schematically in Fig. 3(a), is

$$\rho = \left(\frac{v_{5,6}}{i_{1,4}}\right) \frac{\pi d}{\ln (r_{1,6}\, r_{4,5}) - \ln (r_{1,5}\, r_{4,6})} \tag{5}$$

where $v_{5,6}$ is the potential difference measured between electrodes 5 and 6, $i_{1,4}$ is the current applied between the appropriate electrodes and the $r_{i,j}$ terms

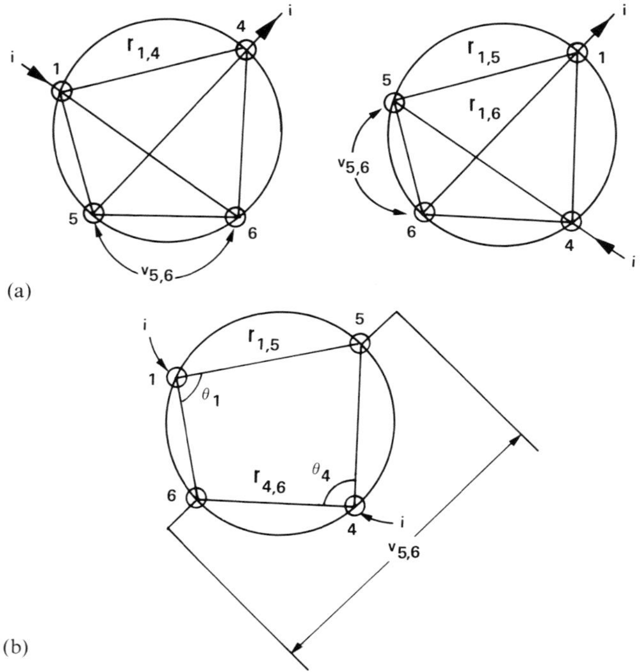


Fig. 3. (a) Commutation of current electrodes for "Corbino image" analysis (eqn. (5)); $r_{i,j}$ are interelectrode distances for resistivity measurement. (b) Current electrodes and potential probes for Hall measurements showing the interelectrode geometry.

represent the actual distances between the corresponding electrodes in Fig. 3(a). For an electrode disposition such as that shown in Fig. 3(b), with an applied transverse magnetic field, the potential difference between electrodes 5 and 6 is

$$v_{5,6} = \left(\frac{\rho i_{1,4}}{\pi d}\right) \ln \left(\frac{r_{1,6}\, r_{4,5}}{r_{1,5}\, r_{4,6}}\right) + \frac{R_h i H}{\pi d} (\theta_1 + \theta_4) \tag{6}$$

where the angles $\theta_1 + \theta_4 = \pi$. If two of the electrodes are on a symmetry axis, which may be a diameter of the circular specimen in Fig. 3(a), and the other two peripheral electrodes are on a perpendicular chord, then geometrical considerations require that $r_{1,6}\, r_{4,5} = 2 r_{1,5}\, r_{4,6}$. Equation (5) is then reduced to eqn. (2) and eqn. (6) to eqn. (3); thus van der Pauw's relations apply in this instance as well.

The voltage measured between the peripheral potential probes of a square or circular specimen with equispaced point-contact electrodes should be independent of the commutation of the current electrodes; therefore the resistance $R_{AB,\,CD} = R_{BC,\,DA} = R$ and eqn. (2) should apply. In practice, if $R_{AB,\,CD} \neq R_{BC,\,DA}$ then the resistivity might be a tensor rather than a scalar quantity. Alternatively, the specimen might be inhomogeneous, containing one or more impurity concentration gradients or a random dispersion of two or more materials or phases. For materials with a ratio $R_0 = R_{AB,\,CD}/R_{BC,\,DA}$ other than unity, the solution to eqn. (1) required for determining the average resistivity $\bar{\rho}$ is[1, 2]

$$\bar{\rho} = \frac{\pi d}{\ln 2} \left(\frac{R_{AB,\,CD} + R_{BC,\,DA}}{2}\right) f(R_0) \tag{7}$$

The functional dependence $f(R_0)$ *versus* $R_0$ has been calculated and presented graphically by van der Pauw. However, it is only partly accurate, with a maximum error of about 2%. A more accurate procedure, described by Price[7], is to use Newton's iterative method to solve eqn. (1). If $\rho_0$ is considered to be an initial trial solution then successive approximations are produced by the iteration

$$\rho_{n+1} = \rho_n - \frac{\rho_n^2 (1 + \alpha + \beta)}{\pi d (\alpha R_{AB,\,CD} + \beta R_{BC,\,DA})} \tag{8}$$

where $\alpha = -\pi d R_{AB,\,CD}/\rho_n$, $\beta = -\pi d R_{BC,\,DA}/\rho_n$ and $\rho_1 = \rho_0 - f(\rho_0)/f'(\rho_0)$ with $f'(\rho_0)$ denoting a derivative with respect to $\rho_0$. Four iterations are usually sufficient for calculating $\bar{\rho}$ with an accuracy to four significant figures.

## 4. SPATIAL INHOMOGENEITIES

Precipitates or inclusions contained in a thin film or layer matrix can have an appreciable effect on the measured $\rho$ and $R_h$ values if the carrier concentrations and mobilities of the inclusions are much larger or much smaller than those of the matrix. If the material is inhomogeneous but the perturbation is relatively minor, if the specimen geometry is simple and if the peripheral electrode disposition is symmetrical, then the method of conformal transformation can be applied for calculating its mean resistivity $\bar{\rho}$. Amer[8] has shown that the mean resistivity of a circular lamina of radius $r$ having a resistivity $\bar{\rho} = \rho_0 + Mr^2$ with $\rho_0 \gg Mr^2$ can be derived from an expression similar to eqn. (1) with $\rho$ replaced by

$$\bar{\rho} = \rho_0 + Mr_0^2/2 \tag{9}$$

This is the same as the average resistivity obtained by integrating $\rho$ over the radius of the lamina:

$$\bar{\rho} = \frac{1}{\pi r_0^2} \int_0^{r_0} 2\pi \rho r \mathrm{d}r = \rho_0 + \frac{Mr_0^2}{2} \tag{10}$$

Amer suggested that similar calculations can be made on inhomogeneous specimens with simple geometrical contours but different resistivity distributions provided that their peripheries are lines of constant resistivity.

Thin films and layers may contain non-conductive inclusions in the form of pinholes or voids which can also affect their electrical and galvanomagnetic properties. Juretschke and Landauer[9] have developed a theoretical treatment which relates the measured $\rho$ and $R_h$ of such inhomogeneous films to those of the same but homogeneous materials and Goldin and Juretschke[10] have confirmed the theoretical analysis experimentally. If $\varepsilon$ is the fractional film volume occupied by the inclusions, which are assumed to have a spherical shape, then

$$\rho_m = \frac{1+\varepsilon/2}{1-\varepsilon} \rho_0 \qquad R_{hm} = \frac{1-\varepsilon/4}{1-\varepsilon} R_{h0} \tag{11}$$

where $\rho_0$ and $R_{h0}$ refer to the homogeneous films. For randomly oriented cylinders, they found that

$$\rho_m = \frac{1+\varepsilon/3}{(1-\varepsilon)(1-\varepsilon/3)} \rho_0 \qquad R_{hm} = \frac{1-2\varepsilon/3}{(1-\varepsilon)(1-\varepsilon/3)} R_{h0} \tag{12}$$

However, if the cylindrical cavities are oriented preferentially along the same direction as that of the applied magnetic field, then the inclusions produce a large change in the measured $\rho$ and only a small or negligible change in $R_h$ with respect to the same parameters of a homogeneous specimen:

$$\rho_m = \frac{1+\varepsilon}{1-\varepsilon} \rho_0 \qquad R_{hm} \simeq R_{h0} \tag{13}$$

van Daal[6] investigated experimentally the errors introduced by conductive and non-conductive inclusions in $SnO_2$ layers 0.3 μm thick deposited onto insulating substrates. For this purpose he used circular specimens with peripheral equispaced electrodes. The conductive inclusions consisted of various configurations of vacuum-deposited circular dots or stripes. Non-conductive inclusions were produced by etching similar patterns down to the substrates. His measurements indicate that the total area covered by the conductive inclusions rather than the ordering or spatial distribution of the inclusions determines the measured resistivity $\rho_m$. The current streamlines tend to crowd into a conductive inclusion and out of a less conductive matrix. The discrepancy between the Hall coefficients measured on an inhomogeneous specimen and on an identical homogeneous specimen was found to depend on the total area covered by the conductive inclusions as well as their spatial distribution. van Daal found that non-conductive inclusions have only a slight effect on $R_h$ but a large effect on $\rho$, a situation clearly in accord with eqn. (13).

An analytical and experimental investigation of the effects of a single conductive inclusion on measured $\rho$ and $R_h$ values was made by Wolfe and Stillman[11] and Wolfe *et al.*[12] They considered the electrical and galvanomagnetic properties of a specimen with a concentric conductive inhomogeneity separated by a sharp boundary from the rest of the cylindrical specimen, as shown in Fig. 4. If the

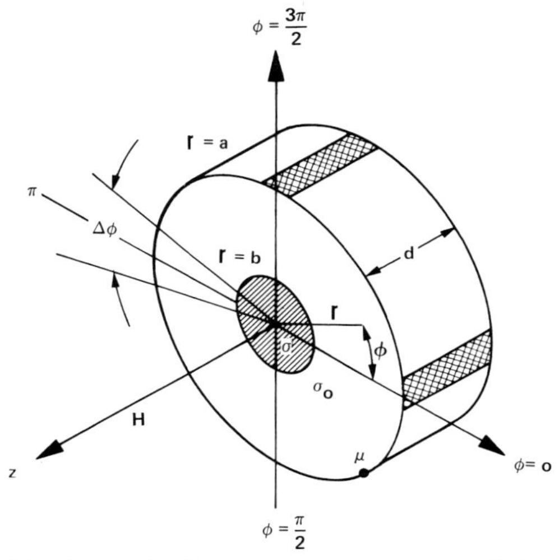


Fig. 4. van der Pauw measurements on a cylindrical specimen with a controlled inhomogeneity, of conductivity $\sigma$ and mobility $\mu$, embedded in a medium of conductivity $\sigma_0$ and mobility $\mu_0$; in cylindrical coordinates with electrodes at 90° positions.

conductivity of the central region is $\sigma$ and that of the rest of the specimen $\sigma_0$ and if $\sigma \gg \sigma_0$ then the inner region can be considered to be an equipotential zone. The current is applied between the contacts at $\phi = \pi/2$ and $\phi = 0$ and the potential difference is measured between the electrodes at $\phi = \pi$ and $\phi = 3\pi/2$. With the boundary condition for the radial current density

$$J_r(a, \phi) = (i/da)\{\delta(\phi - \pi/2) - \delta(\phi - 0)\} \tag{14}$$

where $\delta(\phi - \phi_0)$ is the Dirac delta function and $\phi_0$ the angular position of the contact, the voltage measured between the potential probe is

$$V(a, \pi) - V(a, 3\pi/2) = \frac{i}{\pi d\sigma_0} \sum_{n=1}^{\infty} \frac{(1-\alpha^{2n})^3}{(1+\alpha^{4n})(1+\alpha^{2n})} \frac{(-1)^{n+1}}{n} \tag{15}$$

where the ratio of the radii $\alpha = b/a$; the measured resistivity is

$$\rho_m = \frac{1}{\sigma_0 \ln 2} \sum_{n=1}^{\infty} \frac{(1-\alpha^{2n})^3}{(1+\alpha^{4n})(1+\alpha^{2n})} \frac{(-1)^{n+1}}{n} \tag{16}$$

If the specimen is homogeneous then, evidently, $\alpha = 0$ and the summation in eqn. (16) reduces to ln 2; hence $\rho_m = 1/\sigma_0$. If $\sigma$ is not much larger than $\sigma_0$ then $\rho_m$ depends on the ratio $\sigma/\sigma_0$ as well as on $\alpha$. In particular, if the resistivity of the homogeneous specimen is $\rho_0$ and that of the inhomogeneous specimen is $\rho_m$ then $\rho_m/\rho_0$ decreases with increasing $\alpha$ and $\sigma/\sigma_0$. However, even for small values of $\alpha$ there is an appreciable dependence of $\rho_m/\rho_0$ on $\sigma/\sigma_0$ for $1 < \sigma/\sigma_0 < 10$. Thus relatively small changes in $\sigma/\sigma_0$ can produce significant changes in $\rho_m/\rho_0$ even if the fractional volume of the inhomogeneity is small.

The Hall coefficient of the inhomogeneous specimen in Fig. 4 can be determined by applying a current $i$ between the electrodes located at $\pi/2$ and $3\pi/2$. With the boundary condition for the radial current density

$$J_r(a, \phi) = (i/ad)\{\delta(\phi - \pi/2) - \delta(\phi - 3\pi/2)\} \tag{17}$$

the voltage measured between the potential probes is

$$V(a, \pi) - V(a, 0) = \frac{4i\mu H}{\pi d \sigma_0} \sum_{n=1}^{\infty} \frac{1+(\mu H)^2}{\{(1+\alpha^{4n-2})/(1-\alpha^{4n-2})\}^2 + (\mu H)^2} \frac{(-1)^{n+1}}{2n-1} \tag{18}$$

where $\mu$ represents the charge carrier mobility of the homogeneous portion of the specimen and the Hall coefficient is

$$R_{\text{hm}} = \frac{4\mu}{\pi \sigma_0} \sum_{n=1}^{\infty} \frac{1+(\mu H)^2}{\{(1+\alpha^{4n-2})/(1-\alpha^{4n-2})\}^2 + (\mu H)^2} \frac{(-1)^{n+1}}{2n-1} \tag{19}$$

For $\alpha = 0$ the summation of eqn. (19) reduces to $\pi/4$ and $R_{\text{hm}} = \mu/\sigma_0$. For $\mu H \ll 1$, the measured Hall coefficient $R_{\text{hm}}$ decreases with increasing $\alpha$. The measured resistivity $\rho_{\text{m}}$ also decreases with increasing $\alpha$, in accordance with eqn. (16). However, the decrease of $\rho_{\text{m}}$ with $\alpha$ can be considerably greater than that of $R_{\text{hm}}$. Therefore the apparent mobility calculated from the measured parameters $\rho_{\text{m}}$ and $R_{\text{hm}}$, $\mu_{\text{m}} = R_{\text{hm}}/\rho_{\text{m}}$, can be as much as one order of magnitude larger than $\mu$. For a given value of $\alpha$, as $H$ is increased $R_{\text{hm}}$ increases with $H$ and, since $\rho_{\text{m}}$ is measured in $H = 0$, $\mu_{\text{m}}$ also increases. For $(\mu H)^2 \gg 1$ the measured Hall coefficient approaches $\mu/\sigma_0$, the Hall coefficient of the homogeneous portion of the specimen. Physically, this implies that in the high field limit the current streamlines are distorted out of the inhomogeneous central portion of the specimen and current flow takes place only in the homogeneous region. The experimental measurements made by Wolfe *et al.*[13] on thin epitaxial GaAs layers deposited onto semi-insulating GaAs substrates, with metallic-alloyed gallium inclusions simulating the inhomogeneity, indicate good agreement between theory and experiment. For example, for $\alpha = 0.5$, $\mu_m/\mu$ is nearly independent of $\mu H$ for $\mu H$ values between 0.01 and 1, and increases by a factor of 6 as $\mu H$ increases from 1 to 10. These results suggest that the electron mobility of bulk and epilayer semiconductors used as a material quality index should be treated with caution unless additional information is made available on their homogeneity. Anomalously high electron mobilities, in excess of the theoretical lattice scattering mobility, have been measured on some GaAs epilayers by Wolfe *et al.*[13] They found that these layers contained a random dispersion of conductive precipitates approximately $10^9$ $cm^{-3}$ in concentration. They concluded that anomalously high electron mobilities are likely to be measured in epilayers containing conductive inclusions if (a) the apparent mobilities calculated from resistivity and Hall measurements exceed theoretical expectations, (b) the measurements from different portions of the same wafer are not reproducible, (c) the Hall coefficients have an anomalous magnetic field dependence and (d) the degree of calculated compensation appears to be unusually low.

An extensive literature survey on the conductivities of inhomogeneous materials has been compiled by Reynolds and Hough[14]. Their review is primarily concerned with the properties of a material embedded in another material and separated from it by distinct boundaries. The conductivities of materials containing weak but continuously variable fluctuations in electrical and galvanomagnetic properties have been treated theoretically by Herring[15]. He considered, in particular, the case of inhomogeneities whose scale is small with respect to the size of a specimen but large compared with a Debye length or with the mean free path of charge carriers, and derived the effective resistivity and Hall coefficient in terms of the mean and the local fluctuation about the mean of these parameters for randomly distributed inclusions. He also derived expressions for $\rho$ and $R_h$ of inhomogeneous materials containing spatially ordered inclusions, such as those of a stratified medium with periodically varying carrier concentrations. Such a stratified medium can have a large magnetoresistance[16] in a transverse magnetic field, produced by the distortion of the current streamlines and the short-circuiting of the Hall field by conductive inclusions. Such magnetoresistance can produce large non-linearities in $v_h$ *versus* $H$ and consequently can lead to errors in the calculation of $R_h$. Furthermore, the magnetic field dependence of the magnetoresistance which should saturate in high magnetic fields for homogeneous specimens tends to increase indefinitely with $H$ in inhomogeneous specimens containing conductive inclusions[15]. Beer[17] indicates that for statistically isotropic fluctuations in carrier concentration their effect on the measured galvanomagnetic parameters increases with the tangent of the Hall angle $R_h H/\rho$. However, if the charge carrier fluctuations are limited to directions normal to the applied transverse magnetic field then their influence is even greater; the magnetoresistance, for example, is a quadratic function of $\tan\theta$. Effects such as these are of particular significance in two-phase systems such as InSb–NiSb, in which an ordered array of NiSb needle-like inclusions[18] is grown in the bulk InSb or in inhomogeneous InSb–In layers which consist of an InSb matrix containing ordered filamentary In inclusions[19].

## 5. INHOMOGENEITIES IN THE THICKNESS DIMENSION

Modern semiconductor device technology is based, in part, on the synthesis and growth of epitaxial layers and thin films and on the diffusion or ion implantation of one or more impurity species under controlled conditions. Neither the substrate–epilayer interface nor the diffusion or implantation boundary is actually abrupt. It is of some importance, therefore, to determine gradients in the fundamental charge carrier transport parameters in a direction normal to film or epilayer surfaces. Among the techniques available for the evaluation of such gradients, the measurement of the thickness dependence of $\rho$ and $R_h$ represents one of the least ambiguous approaches.

If a film or epilayer is oriented with its major surfaces in the $xy$-plane of a cartesian system then, considering it to be homogeneous in this plane, it will be assumed to have a $z$-dependent carrier concentration $n(z)$ and mobility $\mu(z)$. Resistivity measurements using van der Pauw's technique were employed by Tufte[20] to determine the surface carrier density and by Subashchiev and Poltinnikov[21], using the van der Pauw Hall configuration, for determining the carrier densities and mobility profiles of diffused layers in silicon.

These methods were also applied to the analysis of planar Zn-diffused InAs layers by Buehler[22], of ion-implanted column III and V dopants in Si by Baron *et al.*[23] and were also investigated by Hlasnik[24] and Pavlov[25]. In order to derive the effective $z$-dependent carrier concentration $n_e(z) = 1/eR_h(z)$ and the effective $z$-dependent mobility $\mu_e(z) = R_h(z)\sigma(z)$, the analytical expressions of Petritz[26] are used for the sheet conductivity $\sigma_s$ and the sheet Hall coefficient $R_{hs}$ respectively:

$$\sigma_s = \sigma d = \int_0^d \sigma(z)\,dz \qquad R_{hs} = \frac{R_h}{d} = \int_0^d \frac{R_h(z)\,\sigma^2(z)\,dz}{\sigma_s^2} \tag{20}$$

The derivatives with respect to $z$ of eqns. (20) lead to

$$\begin{aligned} \sigma(z) &= d\sigma_s/dz \\ n_e(z) &= \frac{\sigma^2(z)}{e\{d/dz(R_{hs}\sigma_s^2)\}} \\ \mu_e(z) &= \frac{1}{\sigma(z)}\frac{d}{dz}(R_{hs}\sigma_s^2) \end{aligned} \tag{21}$$

The experimental procedure consists of sequentially removing incremental thickness layers from the specimen surface, re-measuring $\sigma_s$ and $R_{hs}$ in each case using van der Pauw's technique and evaluating $n_e(z)$ and $\mu_e(z)$ from eqns. (21) by reducing the derivatives to differentials:

$$\frac{\Delta\sigma_s}{\Delta z} = \frac{d\sigma_s}{dz} \qquad \frac{\Delta(R_s\sigma_s^2)}{\Delta z} = \frac{d(R_s\sigma_s^2)}{dz} \tag{22}$$

where $\Delta z$ is the thickness of the layer removed from the specimen and $\Delta\sigma$ and $\Delta(R_{hs}\sigma_s^2)$ are the changes in these quantities produced by the layer removal. The incremental thickness of the layer removed from the specimen needs to be known with fair accuracy. This is usually accomplished by first anodizing and then stripping the anodized layer, or by chemical dissolution or sputter-etching[27, 28], with $\Delta z$ predetermined by controlling the parameters of the anodization or chemical dissolution process; alternatively the step in thickness between an unaffected control pedestal and the etched area of the specimen is measured by appropriate interferometric techniques or by means of a thickness gauge[19]. Methods employing anodization and stripping or etching for determining the thickness dependence and impurity profile of $n$ and $\mu$ suffer from the fact that the changes in these parameters with layer removal appear as small differences taken between large values and are therefore subject to large errors. A different measurement technique for determining $n$ and $\mu$ profiles is based on the use of a Schottky barrier gate on a specimen with a van der Pauw geometry, as shown in Fig. 5. The basic measurement technique has been described in detail by Tansley[29]. From eqn. (7) with $\bar{\rho} = 1/\sigma$, the incremental change in the sheet conductivity $\Delta\sigma_s = \Delta\sigma d$ is

$$\Delta\sigma_s = \frac{2i\ln 2}{\pi f(R_0)} \frac{\Delta V_{AB, CD} + \Delta V_{BC, DA}}{(V_{AB, CD} + V_{BC, DA})^2} \tag{23}$$

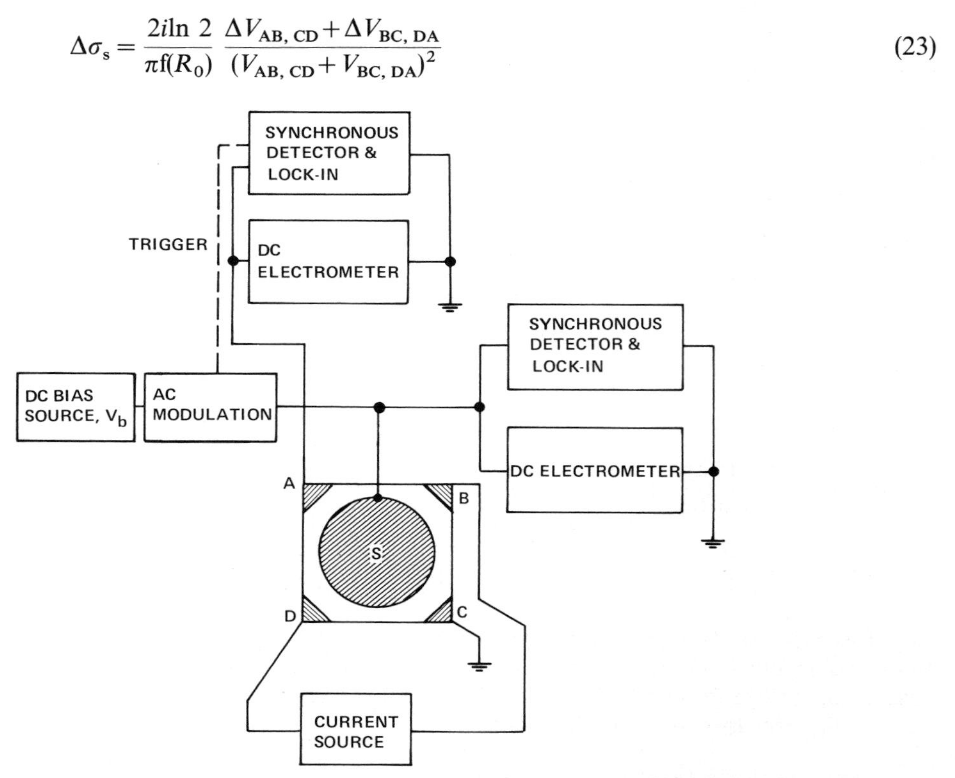


Fig. 5. Schematic diagram of Schottky-gated van der Pauw geometry measurements for determining impurity and mobility profiles of films or epilayers.

Then, on taking the appropriate differentials in eqns. (21) and (22)

$$n_e(z) = \frac{iH}{e\Delta z} \frac{t^2}{\Delta v_h - 2tv_h}$$

$$\mu_e(z) = \frac{2 \ln 2}{\pi H \, f(R_0)} \frac{\Delta v_h - 2tv_h}{\Delta V_{AB, CD} + \Delta V_{BC, DA}} \quad (24)$$

where

$$t = \frac{\Delta V_{AB, CD} + \Delta V_{BC, DA}}{V_{AB, CD} + V_{BC, DA}}$$

Equations (23) and (24) are used in conjunction with the equipment shown schematically in Fig. 5 in the following manner. A fixed reverse bias voltage $V_b$ is applied to the gate; this determines the effective depth $\Delta z$ below the surface in which the specimen is depleted of charge carriers. The superimposed low level a.c. modulation on $V_b$ provides the incremental signals which are demodulated and detected synchronously. The d.c. electrometer is intended to have a sufficiently slow response to be unaffected by the modulation signals detected by the synchronous detection system. Commutation between the appropriate electrodes provides the quantities $V_{AB, CD}$, $V_{BC, DA}$ and $v_h$ and the modulation in the resistivity and Hall voltages provides $\Delta V_{AB, CA}$, $\Delta V_{BC, DA}$ and $\Delta v_h$. The effective thickness $\Delta z$ is determined from capacitance *versus* voltage measurements and the relation $\Delta z = \kappa \varepsilon_0 A/C$, where $\kappa$ is the dielectric constant, $\varepsilon_0$ the permittivity of free space, $A$ the gate area and $C$ the capacitance measured for a specific bias voltage $V_b$. Tansley[29] claims high accuracy and resolution for this method; furthermore 1 % meter readings are adequate for a 5 % error in $n_e(z)$; for the same error the stepwise stripping method requires meter readings to be accurate to 0.01 %. The gated van der Pauw measurements suffer from the same limitation associated with $C$–$V$ measurements: the electric field must be limited below the critical breakdown field. This sets the limit for the ionized impurity density in Si to approximately $2.5 \times 10^{12}$ cm$^{-2}$, which corresponds to a maximum depletion depth of approximately 10 μm for material with a resistivity of about 2 Ω cm but a depth of only $10^2$ Å for a material with a resistivity of about $10^{-2}$ Ω cm.

Similar techniques were used by Ipri[30, 31] and Ipri and Zemel[32] to determine the thickness dependence of the mobility, doping concentration and impurity level distribution of silicon approximately 1 μm thick on sapphire films with a doping concentration of $(2\text{–}6) \times 10^{16}$ cm$^{-3}$. Although the structure used for the experimental investigations employed a six-arm bridge circuit in a "metal–oxide–semiconductor Hall bar" configuration, the same criteria described for the Schottky gate apply here as well and similar considerations also apply to the MIS transistor structure employing silicon on sapphire films described by Elliot and Anderson[33].

## 6. ANISOTROPY

The measurements of $\rho$ and $R_h$ described so far are based, in part, on the implicit assumption that both of these parameters are scalar quantities. However, if the material under consideration has a crystalline structure without elements of symmetry then $\rho$ is a tensor with six components and $R_h$ is a tensor with nine components. van der Pauw[34] has shown that the components of the resistivity tensor can be determined by using six plane-parallel specimens of arbitrary shape but oriented so that low index planes are perpendicular to the lines with direction cosines (100), (010), (001), (011), (101) and (110). For determining the Hall coefficient tensor components three specimens with (100), (010) and (001) orientations are adequate provided that each is measured for three different orientations of the applied magnetic field. Such techniques are hardly applicable to thin films or epitaxial layers, which are usually grown along preferred crystallographic orientations that promote the growth of homogeneous single-crystal layers. However, for an anistropic film or layer characterized by the tensor components $\rho_x$, $\rho_y$ and $\rho_z$ and provided that $\rho_z$ is the component perpendicular to the plane of the layer, then eqn. (7) yields the geometric mean $(\rho_x\rho_y)^{1/2}$ of the other two components of the tensor. This has been shown by Hornstra and van der Pauw[35] and by Price[36] who indicated that three measurements of different specimens each oriented with the $x$-, $y$- or $z$-axes normal to the specimen plane provide the necessary information for determining the respective tensor components. The components of the resistivity tensor in the plane of the specimen can be determined from a rectangular specimen cut in such a manner that its edges are parallel to the principal axes of the resistivity tensor, as shown by Price[7]. Line contacts must be made to the corners of the rectangle whose sides have the dimensions $a$ and $b$ respectively. From the solution of the potential distribution in an anisotropic rectangle he derived the relation

$$\left(\frac{\rho_x}{\rho_y}\right)^{1/2} = -\frac{b}{a\pi} \ln \tanh \left\{\frac{\pi d R_{\mathrm{BC,\,DA}}}{16\,(\rho_x\rho_y)^{1/2}}\right\} \tag{25}$$

where the geometric mean $(\rho_x\rho_y)^{1/2}$ is obtained from the relation

$$\exp\left\{-\frac{\pi d R_{\mathrm{AB,\,CD}}}{(\rho_x\rho_y)^{1/2}}\right\} + \exp\left\{-\frac{\pi d R_{\mathrm{BC,\,DA}}}{(\rho_x\rho_y)^{1/2}}\right\} = 1 \tag{26}$$

and in-plane components of the resistivity tensor are therefore

$$\rho_x = (\rho_x\rho_y)^{1/2}\,(\rho_x/\rho_y)^{1/2} \qquad \rho_y = (\rho_x\rho_y)^{1/2}/(\rho_x/\rho_y)^{1/2} \tag{27}$$

## 7. DISCUSSION

A knowledge of the charge carrier transport parameters of films and epitaxial layers is required for both scientific and technological applications. An analysis of the temperature dependence of the resistivities and Hall coefficients of such layers yields many of these parameters, such as the fundamental energy band gap of semiconductors, the charge carrier concentrations, their mobilities, the impurity doping concentrations and their ionization energies.

The measurements described in this paper are concerned primarily with the application of van der Pauw's technique and its variations to square and circular symmetry specimens. These are considered to be more appropriate for thin films or epilayers than the standard methods of measurements used on bulk crystalline materials. Standard resistivity and Hall measurements[37] can be made on bridge-shaped samples whose length-to-width ratio is chosen to minimize "geometrical magnetoresistance" contributions. A detailed analysis of a double-cross bridge-shaped specimen was given by Jandl *et al.*[38] who showed that if the proper geometrical limitations on the size and spacing of the lateral arms are observed then the electrodes can be considered as point contacts and the potential difference measured between the bridge arms can be within 1% of that between two point contacts on the main body of the specimen. The four-point probe method[39] is a standard procedure for measuring the resistivity of silicon wafers. In its simplest form it consists of four in-line equispaced tungsten point contacts applied, under some pressure, to the surface of a wafer. If the measurements are performed in a region remote from the wafer boundaries then the resistivity can be represented[40] by eqn. (2) multiplied by a correction factor which is a function of the ratio of the wafer thickness to interprobe spacing $s$:

$$\rho = (\pi d R / \ln 2)\ \mathrm{F}(d/s) \tag{28}$$

The correction factor $\mathrm{F}(d/s)$ has been tabulated by Uhlir[41]. Four-point probes used in a square array can be used for measuring resistivity as well as Hall coefficients, as shown by Buehler and Pearson[3] and by Buehler[4]. However, due account must be taken of the specimen geometry by means of appropriate correction factors. A variety of errors which may be cumulative are encountered with pressure-type four-point probe contacts. These errors include displacement of some of the contacts due to uneven pressure executed upon them, as investigated by Hall[42], minority carrier injection at surface barriers, particularly in high band gap semiconductors and at low temperatures, proximity of the specimen boundaries, dislocations introduced into specimens by too high a contact pressure and heating at the contacts by excessive input currents. Peripheral symmetrical "ohmic" electrodes applied to a specimen of simple geometry, as described in this paper, have therefore considerable advantages for measuring $\rho$ and $R_h$. However, such measurements represent averages over a specimen; other techniques, outside the scope of this paper, have to be used for identifying and determining quantitatively localized fluctuations in carrier concentration such as those produced by localized impurity concentration gradients, individual inclusions or precipitates.

REFERENCES

1 L. J. van der Pauw, *Philips Res. Rep.*, *13* (1958) 1.
2 L. J. van der Pauw, *Philips Tech. Rev.*, *20* (1958) 220.
3 M. G. Buehler and G. L. Pearson, *Solid State Electron.*, *9* (1966) 395.

4 M. G. Buehler, *Solid State Electron.*, *10* (1967) 801.
5 R. Chwang, B. J. Smith and C. R. Crowell, *Solid State Electron.*, *17* (1974) 1217.
6 H. J. van Daal, *Philips Res. Rep. (Suppl.)*, No. 3 (1965) 8–14.
7 W. L. V. Price, *Solid State Electron.*, *16* (1973) 753.
8 S. Amer, *Solid State Electron.*, *6* (1963) 141.
9 H. J. Juretschke and R. Landauer, *J. Appl. Phys.*, *27* (1956) 838.
10 E. Goldin and H. J. Juretschke, *Trans. Metall. Soc. AIME*, *212* (1958) 357.
11 C. M. Wolfe and G. E. Stillman, *Appl. Phys. Lett.*, *18* (1971) 205.
12 C. M. Wolfe, G. E. Stillman and J. A. Rossi, *J. Electrochem. Soc.*, *119* (1972) 250.
13 C. M. Wolfe, G. E. Stillman, D. L. Spears, D. E. Hill and F. V. Williams, *J. Appl. Phys.*, *44* (1973) 732.
14 J. A. Reynolds and J. M. Hough, *Proc. Phys. Soc. London*, *B70* (1967) 769.
15 C. Herring, *J. Appl. Phys.*, *31* (1960) 1939.
16 H. Weiss, 1, Physics of III–V compounds. In R. K. Willardson and A. C. Beer (eds.), *Semiconductors and Semimetals*, Academic Press, New York, 1966, pp. 315–376.
17 A. C. Beer, *Galvanomagnetic Effects in Semiconductors*, Academic Press, New York, 1963, pp. 308–328.
18 H. Weiss *Structure and Application of Galvanomagnetic Devices*, Pergamon Press, Oxford, 1969, pp. 63–80.
19 H. H. Wieder, *Intermetallic Semiconducting Films*, Pergamon Press, Oxford, 1970.
20 O. N. Tufte, *J. Electrochem. Soc.*, *109* (1962) 235.
21 V. K. Subashchiev and S. A. Poltinnikov, *Sov. Phys. Solid State*, *2* (1960) 1059.
22 M. G. Buehler, *Stanford Res. Rep.*, *SEL-66-064*, 1966.
23 R. Baron, G. A. Shifrin, O. J. Marsh and J. W. Mayer, *J. Appl. Phys.*, *40* (1969) 3702.
24 I. Hlasnik, *Solid State Electron.*, *8* (1965) 461.
25 N. I. Pavlov, *Sov. Phys. Semicond.*, *4* (1971) 1644.
26 R. L. Petritz, *Phys. Rev.*, *110* (1958) 1254.
27 S. M. Davidson, *Proc. 2nd Int. Conf. on Ion Implantation, Garmisch-Patenkirchen*, Vol. 2, Springer Verlag, Berlin, 1971, p. 79.
28 O. Cahen and B. Netange, *Proc. Europ. Conf. on Ion Implantation, Reading*, Peter Peregrinus, Stevenage, 1970, p. 192.
29 T. L. Tansley, *J. Phys., E, 8* (1975) 52.
30 A. C. Ipri, *Appl. Phys. Lett.*, *20* (1972) 1.
31 A. C. Ipri, *J. Appl. Phys.*, *43* (1972) 2770.
32 A. C. Ipri and J. N. Zemel, *J. Appl. Phys.*, *44* (1973) 744.
33 A. B. M. Elliot and J. C. Anderson, *Solid State Electron.*, *15* (1972) 531.
34 L. J. van der Pauw, *Philips Res. Rep.*, *16* (1961) 187.
35 J. Hornstra and L. J. van der Pauw, *J. Electron. Control*, *7* (1959) 169.
36 W. L. V. Price, *J. Phys. D*, *5* (1972) 1127.
37 *Standard Method for Measuring Hall Mobility in Extrinsic Semiconductor Single Crystals*, ASTM F76-68, August 1968.
38 S. Jandl, K. D. Usadel and G. Fischer, *Rev. Sci. Inum.*, *45* (1974) 1479.
39 L. B. Valdes, *Proc. IRE*, *42* (1954) 420.
40 F. M. Smits, *Bell Syst. Tech. J.*, *37* (1958) 711.
41 A. Uhlir, *Bell Syst. Tech. J.*, *34* (1955) 105.
42 R. Hall, *J. Sci. Instrum.*, *44* (1967) 53.

*Thin Solid Films*, *31* (1976) 139–170



# A REVIEW OF ETCHING AND DEFECT CHARACTERISATION OF GALLIUM ARSENIDE SUBSTRATE MATERIAL

D. J. STIRLAND

*The Plessey Company Limited, Allen Clark Research Centre, Caswell, Towcester, Northants. (Gt. Britain)*

B. W. STRAUGHAN

*R.R.E., Great Malvern, Worcs. WR14 3PS (Gt. Britain)*

(Received May 9, 1975; accepted June 17, 1975)

Emphasis is placed on the practical aspects of the etching of GaAs substrates, including procedures for the preparation of polished substrates and for the revelation of defects by selective attack. Basic information on the crystallography of GaAs and its effects on the structures of defects is considered, with particular reference to the reactions of etching solutions at the defects. Problems associated with the introduction of mechanical damage at substrate surfaces and its detection and elimination by various etch treatments is discussed.

Etch compositions for the preparation of various surface finishes are described and classified. Complex etch features resulting from the attack of particular etchants at substrate surfaces are interpreted in terms of defect structures. Evidence is presented which demonstrates that correct surface treatment is necessary prior to the application of certain defect etches for their successful operation.

---

## 1. INTRODUCTION

Etching can be defined quite simply as the removal of surface material. This review considers in some detail chemical etching procedures which have been developed for gallium arsenide substrates. It will exclude other removal methods such as thermal etching and ion sputtering techniques. The aim is not to provide a complete catalogue of all the reported etches for gallium arsenide *per se*, but rather to illustrate via a range of practical examples specific to this material some of the principles underlying etching procedures, mechanisms, and the interpretation of results.

There are several reasons why it may be necessary to remove material from a surface. Featureless and planar surfaces are a prerequisite for the success of semiconductor device processes such as epitaxial growth. Planar surfaces can be produced by various combinations of mechanical and chemical polishing; the final removal of superficial damage and localised irregularities is usually achieved by a chemical polishing solution. A variety of etchants have been devised for this purpose; we shall call these *polish etches* or *free etches*.

Although the surface produced by the action of a polish etch may be smooth and free of superficial damage, inevitably it will intersect some of those crystallographic defects which are distributed throughout the volume of the material. The intersection sites can be revealed by the use of etchants which we shall call *defect etches*.

Faust[1] has drawn a distinction between two types of defect etch. First we have "*preferential*" etchants which produce etch pits which are faceted and the surrounding surface does not necessarily have a polished finish. The faceted pits provide a means of orienting the surface by optical goniometry. The second type of defect etchant described by Faust is termed "*non-preferential*", and in this case dislocation etch pits are produced which are not faceted and the surrounding surface is polished.

## 2. CHEMISTRY OF ETCHING

The etching process involves electron transfer between the surface being dissolved and the chemical components of the etch. The object is the removal of gallium and arsenic atoms from the surface in forms which are soluble in the etch solution. The etching rate may bear some relationship to the defects present in the material, its crystallographic orientation and the surface finish[2]. If a feature increases the etch rate, a pit will form at the site; a hillock will form if the feature etches slower than the surrounding matrix[3].

The dissolution rate can be controlled by the diffusion of reactants to the semiconductor surface and the reaction products away from it. The activation energy for this process is usually low and of the order[4] of 5 kcal $mol^{-1}$. Slight variations in the etch composition or lowering the etch temperature can reduce the dissolution rate so that the etch becomes reaction rate controlled. In this case the rate is determined by the chemical activity of the surface[4].

Etches for gallium arsenide are usually oxidising solutions in concentrations which prevent the deposition of arsenious oxide[5,6]. Oxidation is essentially an electrochemical process in which it is considered that localised anodic and cathodic areas exist at the semiconductor surface. The material goes into solution at anodic sites and the oxidising agent is reduced at cathodic sites[7]. It follows that the more anodic an area the faster will be its dissolution and as a consequence an etch pit will form. Any external agency which creates electrons and holes at the surface will affect the etching rate. Good examples of this are the effects of illumination[8] or of applied electric currents[9].

It is also possible to observe different etching effects by altering the ratio of components of a particular etch, as demonstrated by Mullin *et al.*[10] for InP and Straughan[11] for GaAs.

A recent review which includes a detailed discussion of the chemistry of etching processes at semiconductor surfaces has been given by Tuck[12].

## 3. CRYSTALLOGRAPHY OF GALLIUM ARSENIDE

An important etch rate determining factor for gallium arsenide is the crystallographic orientation of the surface and the non-equivalence of $\{111\}$ faces.

GaAs belongs to the cubic non-centrosymmetrical space group $F\bar{4}3m$, *i.e.* it has the sphalerite (zincblende) structure. This is the same as the diamond structure except that two types of atoms are present: each Ga (group III) atom has four As (group V) atoms as nearest neighbours and vice versa. The III–V compound bonding is essentially covalent[13], with $sp^3$ hybridised orbitals in tetrahedral array connecting the two atom species, so that the GaAs crystal structure can be considered as a network of $<111>$ direction bonds linking alternating Ga and As atoms in space. We shall adopt the convention introduced by Gatos and Lavine[14] that group III atoms are designated as A atoms and group V atoms as B atoms.

Figures 1 and 2 show two views of the GaAs crystal lattice, along $<112>$ and $<110>$ directions. Black and white balls represent A- and B-type atoms respectively, and the cubic unit cell is indicated (the lattice parameter of undoped[15] GaAs is $a_0 = 5.65325 \pm 0.00002$ Å at 27 °C).

The atomic arrangement imposes a polarity on the lattice in all crystallographic directions except $<100>$ and $<110>$. This polarity is most clearly

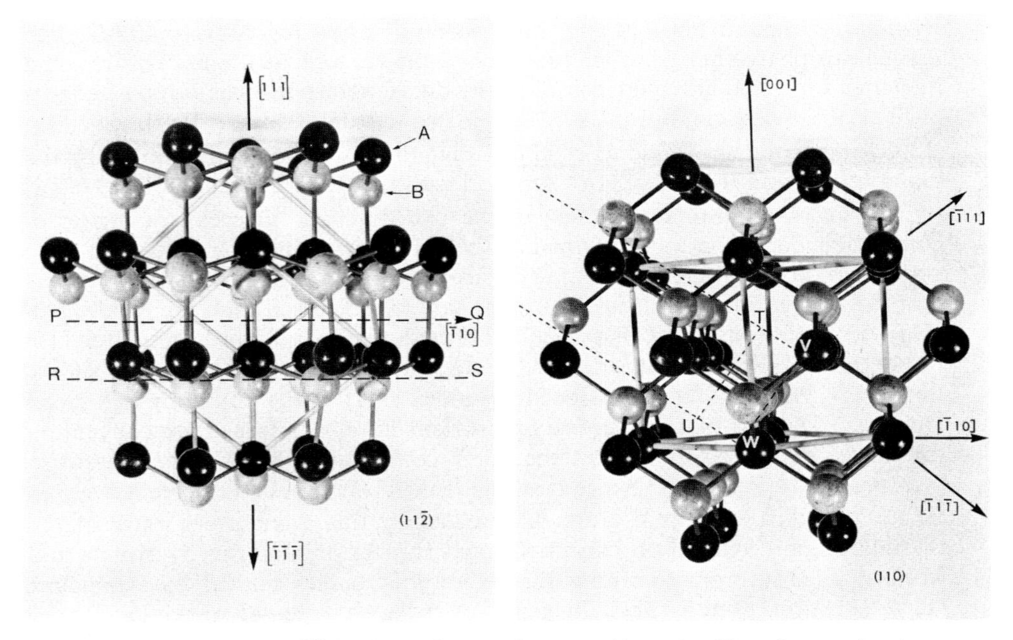


Fig. 1. View of GaAs along $[\bar{1}\bar{1}2]$: black spheres ≡ Ga atoms (A type); white spheres ≡ As atoms (B type).
Fig. 2. View of GaAs along $[\bar{1}\bar{1}0]$.

evident by consideration of the [111] and $[\bar{1}\bar{1}\bar{1}]$ directions in the structure. Along the [111] direction the atoms alternate in the sequence –BA–BA–BA, whereas along $[\bar{1}\bar{1}\bar{1}]$ the sequence is –AB–AB–AB–. The $(\bar{1}\bar{1}\bar{1})$ and (111) planes can be considered as double sheets of (BA) or (AB) atoms, with each atom triply bonded to its neighbours within the double sheet. The bonding between one double sheet and the next double sheet is a single bond per atom (see PQ on Fig. 1), so that it is expected that (111) or $(\bar{1}\bar{1}\bar{1})$ surfaces will terminate at such positions. Surfaces formed by cutting along RS (Fig. 1) would require three broken bonds per atom and are not expected to be stable[14, 16]. Thus the lack of a centre of symmetry in the sphalerite structure manifests itself in the differing chemical structures of (111) and $(\bar{1}\bar{1}\bar{1})$ surfaces. Reference to Fig. 1 shows that the surface formed below PQ consists of A (Ga) atoms and that its crystallographic plane is (111); the surface above PQ consists of B (As) atoms and is a $(\bar{1}\bar{1}\bar{1})$ plane, given the crystallographic directions as specified.

The non-equivalent structures of (111) and $(\bar{1}\bar{1}\bar{1})$ GaAs surfaces are revealed by their different behaviour in a variety of etchants. This will be considered in more detail in later sections, but it is necessary to anticipate some of these results at this point. White and Roth[17] found that opposite faces of {111} GaAs slices were attacked differently in an etchant consisting of 2:1:2 $HCl{:}HNO_3{:}H_2O$. One side, called the A side, gave many etch pits, whereas the opposite face, called the B side, showed dark patches (probably reaction products adhering to the surface). Third-order reflections were recorded from each face in turn, using an X-ray fluorescent target to provide Se K$\alpha$ and Br K$\alpha$ X-radiation. These wavelengths are close to the Ga and As absorption edges respectively and lead to anomalous phase changes when scattered at the Ga and As atoms. This resulted in changes in the relative intensities of the diffracted beams when opposite faces of the crystal were used in place of the normal crystal analyser. By this method White and Roth were able to establish that the A face corresponded with the Ga-rich face, and the B face corresponded with the As-rich face.

The crystallographic labelling of the two different surfaces is a matter of convention; face A can be specified as the (111) surface and face B as the $(\bar{1}\bar{1}\bar{1})$ surface, or vice versa. An interesting discussion by Hulme and Mullin[18] considers that the convention with historical merit should have been universally adopted. (Dewald[19] published the first account of polarity in indium antimonide. He found that anodic oxide films formed more readily on one (111) face than on the opposite $(\bar{1}\bar{1}\bar{1})$ face, and he assigned the index (111) to the antimony-rich face.) However, it is clear that the majority of workers have *not* adopted this convention for GaAs, or indeed for any of the III–V compounds. Most authors seem to favour the {111} A or {111} B face as an adequate labelling system, where A $\equiv$ group III atom and B $\equiv$ group V atom. Unfortunately this is incorrect usage of the crystallographic symbolism expressed by {111}. As pointed out by Hulme and Mullin[18], {111} represents only the set of *four* planes related by symmetry: (111), $(\bar{1}\bar{1}1)$, $(1\bar{1}\bar{1})$ and $(\bar{1}1\bar{1})$. In order to specify the set of opposite polarity planes $(\bar{1}\bar{1}\bar{1})$, $(11\bar{1})$, $(\bar{1}11)$ and $(1\bar{1}1)$, the symbol $\{\bar{1}\bar{1}\bar{1}\}$ must be used. Thus in their view the correct specification of the opposite non-specific faces would be the $\{\bar{1}\bar{1}\bar{1}\}$ A which is gallium-rich, and the {111} B which is arsenic-rich. The convention we shall adopt throughout this review is now in common usage: *opposite faces are specified as {111} Ga and {111} As surfaces.*

## 4. DEFECT STRUCTURES IN GALLIUM ARSENIDE

One of the uses of etching procedures is the revelation of defects present in grown ingots and substrates. This section will describe the defects which are being sought and have been found, and will indicate the importance of the sphalerite structure in relation to particular defects in GaAs.

### 4.1. *Point defects*

The most numerous defects expected to be present in GaAs single crystals are point defects. The compositional range for the intermetallic sphalerite structure phase in GaAs extends[20] from 49.998 to 50.009 at.% As. This corresponds to concentrations of $4 \times 10^{18}$ As vacancies $cm^{-3}$ and $2 \times 10^{19}$ Ga vacancies $cm^{-3}$ respectively[21], assuming that non-stoichiometry results in vacancies only. Potts and Pearson[22] showed by lattice constant measurements on GaAs crystals subjected to high temperature annealing and quenching treatments under various As overpressures that the primary defects observed were vacancies ($10^{19}$ $cm^{-3}$) on As sites, and that significant numbers of Ga vacancies were not present. However, other authors, *e.g.* Brice and King[23], have suggested that Ga vacancies occur in significant concentrations. It is beyond the scope of this paper to discuss the variety of point defect types which have been proposed theoretically; a useful review has been given by Logan and Hurle[24].

It is not possible to detect individual point defects by etching techniques. However, when a grown crystal is cooled from the growth temperature it becomes supersaturated with vacancies if there are no vacancy sinks available in the bulk, and these vacancies may form clusters or aggregates. This has been demonstrated to occur in dislocation-free silicon[25], in which it was proposed that vacancy–oxygen impurity clusters were produced and could be revealed by etching methods. Zhelikhovskaya and Borisova[26] have associated small etch pits with aggregates of vacancies in low dislocation density GaAs crystals. It was proposed that the aggregates were vacancy–copper clusters, the copper having diffused through into the melt from the quartz boat during the crystal growth process.

### 4.2. *Line defects*

#### 4.2.1. *Dislocations*

Hornstra[27] has classified dislocations which may be expected to occur in the diamond lattice. A typical example is the so-called 60° dislocation (No. II in his nomenclature). In this case the axis of the dislocation lies in, say, a $[1\bar{1}0]$ direction, and its Burgers vector ($\boldsymbol{b} = \frac{1}{2} a_0 \, [10\bar{1}]$) is at 60° to the axis. Haasen[28] first indicated that two different configurations of this dislocation are expected to occur in the case of III–V compounds with the sphalerite structure. The core structure of a 60° dislocation contains atoms with dangling or "broken" bonds, and the $[1\bar{1}0]$ axis of the dislocation lies along a set of identical type atoms in GaAs. Thus two forms can occur, as shown in Fig. 3 at (a) and (b), depending on

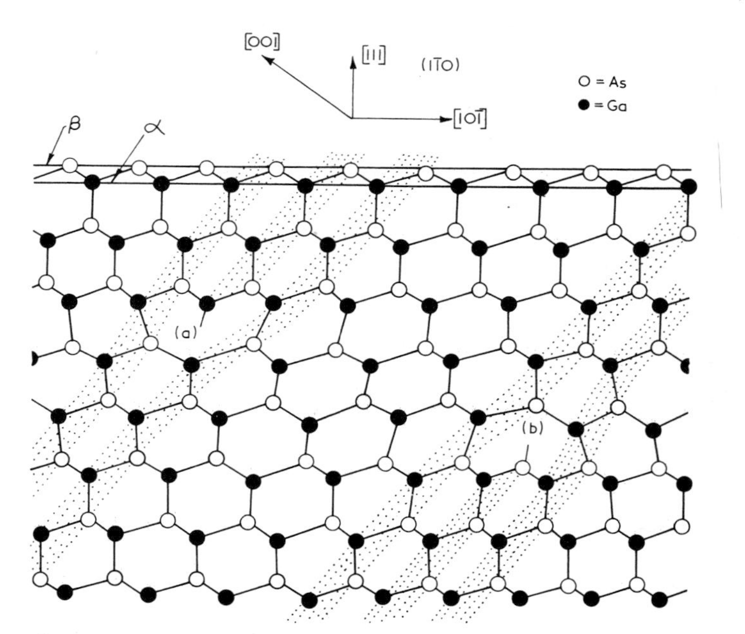


Fig. 3. 60° dislocations, at (a) and (b), in the GaAs lattice.

whether the broken bonds belong to Ga or As atoms. Venables and Broudy[29] named rows of like atoms α for group III atoms and β for group V atoms (see Fig. 3). This notation has subsequently[14, 30] been altered into a nomenclature for the two types of dislocation: the dislocation of Fig. 3 at (a) is called the α or α(Ga) dislocation and that of Fig. 3 at (b) is the β or β(As) dislocation[30]. The dotted lines in Fig. 3 represent a possible choice for the extra (double) half-planes of Ga–As atoms which make up the dislocations, and it can be seen that the α dislocation is a positive dislocation, whereas the β dislocation is negative. Recently Abrahams *et al.*[31] have demonstrated that sets of similar sign α and β 60° dislocations can occur under certain conditions, but the zincblende structure requires that the α set is orthogonal to the β set (and that they lie on different sublattices). This is because equivalent covalent bond directions rotate by 90° on moving from the Ga (or As) sublattice to the As (Ga) sublattice.

Similar considerations due to the zincblende structure of GaAs apply to most of the other dislocations classified by Hornstra[27] and have been dealt with fully by Holt[13].

A further complication is the possible occurrence of two alternative configurations of each dislocation in the diamond lattice. This was first pointed out by Hirthe and Lothe[32], who named the two sets " shuffle " and " glide " dislocations. All the dislocations considered by Hornstra[27] belonged to the " shuffle " set only. Abrahams and his coworkers[31, 33] have indicated that equivalent sets are possible in the zincblende structure. Fundamentally they result from the double-layer atomic arrangement of the zincblende (and diamond) structure shown in Fig. 1. The usual method for visualising the introduction of a dislocation into a crystal structure is to consider the cutting and removal of a half-plane atomic sheet, followed by the rejoining of the crystal. Examination of Fig. 2 shows that the cutting procedure can be accomplished in two different ways for the zincblende structure. Bonds can be severed either along TU or along VW. The former will lead to dislocations of the " shuffle " set, similar to the dislocations shown in Fig. 3, whereas the latter will give dislocations of the " glide " set, shown in Fig. 4. As Olsen *et al.*[33] point out, it is necessary to determine whether α- and β-type dislocations exist as " shuffle " or " glide " sets before consideration can be given to the reasons for different mobilities of the α and β dislocations[34], but this problem has not yet been solved.

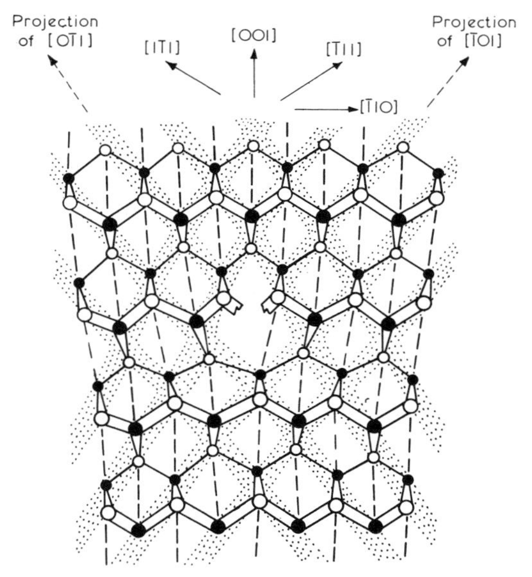


Fig. 4. Possible structure of an edge dislocation in GaAs lattice.

*4.2.2. Dissociated dislocations and stacking faults*

Stacking faults in the diamond lattice can be formed by the aggregation of pairs of vacancies, or pairs of interstitials, to result in either the removal of part of a double {111} atom sheet or the insertion of part of an additional double {111} sheet. The former produces an intrinsic stacking fault, the latter an extrinsic stacking fault. Hornstra[27] has shown that a dislocation in the diamond lattice can lower its energy by dissociating into a strip of faulted material (a stacking fault) which is bounded by two partial dislocations. Holt[13] has considered the modifications imposed on these defects by the zincblende structure. The analysis is similar to that given above for dislocations and shows that in geometric terms two types of fault are expected in the zincblende structure for each fault in the diamond structure. In principle the GaAs faults might be expected to exhibit differing chemical behaviour (*e.g.* etching behaviour) although we know of no experimental evidence to support this view.

*4.3. Volume defects*

*4.3.1. Twinning*

This term describes the sudden change of orientation of a growing crystal where the new growth axis bears a mirror-image relationship to the original axis. If another change occurs afterwards to restore the initial orientation, the mis-oriented section is called a " twin-lamella ". Twinning takes place on a particular crystallographic plane when a fault occurs in the sequence in which the atoms take up lattice positions. The effect is not localised like stacking faults but can extend right across the twin plane.

The twin plane for GaAs is {111}, in common with other III–V compounds. In practice it is found that twinning does not often occur in the growth of GaAs from oriented seeds but if it does take place the specific twinning plane is usually {111} As. This can easily be observed on crystals grown on <001> seeds, because in this case the two polar {111} types intersect the growth interface and twinning occurs only on the As terminating facets. Steinemann and Zimmerli[35] have investigated twinning of GaAs as a function of excess As. They confirm that twinning on the {111}.As face is much more sensitive to As pressure or melt composition compared with that on the {111} Ga face.

*4.3.2. Precipitation*

A number of papers have dealt with the detection and analysis of precipitation phenomena in GaAs, covering a wide range of precipitate compositions, and sizes ranging from hundreds of microns to a few ångströms. The detection method used for larger precipitates has often been infrared transmission microscopy (for a review, see ref. 36) or X-ray topography[37, 38]; the direct determination of the crystallographic structure of smaller precipitates has usually involved transmission electron microscopy in conjunction with electron diffraction analyses. Comprehensive details of these techniques have been given by Hirsch *et al.*[39]

Eckhardt[40] was one of the first authors to determine the composition of precipitates in as-grown undoped GaAs crystals by electron diffraction. The precipitates were $\alpha$-$Ga_2O_3$ crystallites formed by the reaction of GaAs with the quartz containers used in the crystal growth process.

In crystals deliberately doped with Te, precipitates have been identified by electron microscopy as $Ga_2Te_3$ crystallites[41–43]. Detection by etching methods is not possible in all cases, owing to the lower resolution available. In general, etching reveals the locations of the precipitates but cannot identify their composition. Iizuka[37] has described two different types of impurity aggregates which can be revealed by etching heavily Te-doped ($>10^{18}$ $cm^{-3}$) GaAs. Massive triangular-shaped inclusions in both {111} Ga and {111} As faces were found to be less soluble in various etchants, and hence were elevated above the surrounding matrix. Electron probe X-ray microanalysis of the inclusions indicated that they were a Te–Ga complex. A second investigation of similar material[38] showed numerous flat-bottomed saucer-shaped pits after etching. These pits occurred at densities of about $10^5$–$10^7$ $cm^{-2}$, with a maximum diameter of approximately 10 μm, and were associated with micro-precipitates of Te (or Te–Ga complexes). Similar smooth pits were subsequently discovered in pulled {111} GaP crystals[44–46] and were called "s-pite" (saucer pits). As these pits exhibit the same characteristic forms in both GaAs and GaP, we shall use the term s-pit to designate them for both materials. We have found s-pits in (001) GaAs crystals doped with Te, Se or Si to approximately $10^{18}$ $cm^{-3}$ (see Section 5.3.2).

*4.3.3. Precipitates and dislocations*

The stress field around certain types of dislocations can attract impurities to them. This causes an impurity atmosphere to be formed in the vicinity of the dislocations, and if the solubility limit is exceeded the impurity may precipitate out as a separate phase. The dislocations are said to be "decorated" by the impurity, and the effects are often observable by etching. Iizuka[38] found that cylindrical regions around grown-in dislocations in heavily Te-doped $\{111\}$ GaAs were denuded of s-pits. This implied that the dislocations were acting as sinks for the impurities, and it was expected that the dislocations would show an increased impurity content at their cores. Iizuka failed to find this effect with electron probe X-ray microanalysis, probably because of large X-ray absorption in the dislocation etch pit. We have found that impurity decoration of dislocations can be readily observed on (001) etched surfaces of Te-, Cr- and Se-doped GaAs (see Figs. 9–11, Section 6.3).

The duplex nature of some dislocations in the GaAs structure provides the possibility that one of the two types may be decorated differently from the other type by specific impurities. Abrahams and Buiocchi[47] and Abrahams and Pankove[48] have interpreted etching observations on misfit dislocations at p–n junction regions, where the p-type dopant was Zn ($\sim 2\times10^{19}$ $cm^{-3}$). It was proposed that the misfit dislocations, which lay in orthogonal $<110>$ directions, were pure edge dislocations on (001) with Burgers vectors of $\frac{1}{2}a_0[110]$ and $\frac{1}{2}a_0[1\bar{1}0]$. (Later work[49] using electron microscopy to determine Burgers vectors of similar dislocations provided reasonable confirmation of this suggestion.) One set of dislocations, along [110], etched unevenly and showed discrete particles along the lengths of the dislocations, whereas the $[1\bar{1}0]$ set did not. The authors suggested that the [110] set showed enhanced etching due to greater impurity segregation of the zinc dopant than at the $[1\bar{1}0]$ set. Figure 4 shows the postulated structure of the [110] misfit edge dislocation[50]; the orthogonal $[1\bar{1}0]$ dislocation is obtained by interchanging black (Ga) atoms with white (As) atoms. This type of dislocation can be formed by the interaction of two 60° dislocations (shown shaded in Fig. 3) running together on intersecting $\{111\}$ glide planes, *i.e.*

$$\tfrac{1}{2}\,a_0\,[101]+\tfrac{1}{2}\,a_0\,[01\bar{1}]\rightarrow\tfrac{1}{2}\,a_0\,[110]$$

Abrahams and Buiocchi suggest that because Zn will normally substitute for Ga atoms in GaAs then a Zn atom would be attracted to the site between the two incompletely bonded As atoms in Fig. 4. No electronic bonding would be likely to occur for the orthogonal dislocation with incompletely bonded Ga atoms.

*4.3.4. Growth striations and delineation of facets*

Another form of impurity aggregation can take place during the growth of crystals, and it occurs as striations. These are a record of the shape and progress of the freezing interface during crystal growth. Coarse structure in pulled crystals matches the rotation and lift rates during growth, but a fine structure is also present. Hurle[51] has shown that when a melt is subjected to a temperature gradient thermal oscillations are set up, and growth advances in a non-smooth fashion. The reason for this is that convection currents present in the melt bring alternatively cooler and hotter material to the interface. As a consequence, when each growth step is momentarily checked there is a change in the amount of impurity or dopant incorporated in the crystal at that instant. This small change in composition can be detected by suitably selective etches (see Section 5.3.2).

Further complications can occur owing to the "facet effect", as originally observed in InSb by Hulme and Mullin[52]. Almost all added or naturally present impurities have segregation constants $k$ ($= C_s/C_L$) $< 1$, where $C_s$ and $C_L$ represent the concentrations of impurity in the solid and liquid respectively. This value of $k$ can change dramatically to become greater than unity if certain low-index facets are presented at the advancing solid–liquid interface. Pronounced impurity bands parallel to the facet can occur and can be revealed by etching, as reported for GaAs by LeMay[53].

*4.3.5. Cell structure*

When a crystal is growing from a melt which is constitutionally supercooled the advancing growth interface becomes unstable. The near planar interface can become faceted on a microscale, in contrast with the facet effect described above. "Fingers" of solid material grow rapidly into the supercooled melt and between these "fingers" rejected dopant creates a series of extended cell walls enclosing regular-shaped areas of pure material. A more thorough explanation of the effect as demonstrated in germanium has been reported by Bardsley *et al.*[54] Cell structures can be revealed in GaAs by suitable etchants (see Section 5.3.2).

### *4.4. Work damage*

No precise definition of work damage can be given. It is probable that it includes many of the defects described previously in this section, plus additional features such as microcracks and chipped material, local lattice strain and distortion and impurity incorporation.

Damage can be introduced in GaAs by cutting, lapping, chemical/mechanical polishing, scribing, tweezer handling, ion implantation and various device processes such as bonding of contacts, mounting and encapsulation.

It is always essential to remove preparation damage from GaAs substrates if they are subsequently to be used for epitaxial growth or for assessment. In order to effect this removal economically, it is necessary to know the damage depth, which in general will vary considerably with the procedures which have introduced the damage. Tietjen *et al.*[55] used the perfection, or otherwise, of homoepitaxial vapour phase layers as a means of assessing the damage remaining at (001) substrate surfaces after various treatments, and hence could estimate damage depths. They found that the work damage introduced by saw-cutting extended at least 25 μm into the substrate, and that removal of 100 μm (by chemical polishing) provided a nearly defect-free surface.

Some workers have used a mechanical polish after cutting to try and remove the saw damage. Tietjen *et al.*[55] removed 100 μm from sawn substrates with 0.5 to 0.05 μm alumina powders. When a further 25 μm was removed by chemical polishing it was found that pits were produced in subsequently grown epitaxial layers. The authors ascribe these pits to the presence of embedded abrasive particles in the substrate surface due to the mechanical polishing. Removal of 100 μm by chemical polishing eliminated this effect.

Lidbury[56] examined (001) GaAs substrates which had been pad polished with $\frac{1}{4}$ μm diamond paste and he found by electron microscopy examination that removal of 5 μm by chemical polishing was necessary to remove the work damage completely. Removal of 2 μm caused the disappearance of scratches, chips and cracks, but a high density of dislocation tangles was evident. Meieran[57] found that {111} Ga surfaces were less susceptible to 1 μm diamond mechanical polishing damage than {111} As and {001} surfaces. Chemical polishing to remove 20 μm was necessary for 1 μm diamond polished {111} As surfaces.

Steinhardt[58] showed that scratches can act as dislocation loop sources, and Laister and Jenkins[59] found that sapphire-scribed lines on {111} Ga surfaces produced loops which moved at least 50 μm into slices after annealing treatments at 1040 °C. Abrahams and Ekstrom[50] have demonstrated that indentation by a conical diamond can produce limited dislocation movement (plastic deformation) at room temperature.

Many of the etch compositions which are described in the following section can be utilised to reveal work damage, but not all etches are sensitive to work damage. Abrahams and Ekstrom[60] examined a metallographically polished {111} Ga surface which had been given a final polish with 1 μm diamond paste on a cloth-covered wheel. The authors state, " when examined at a magnification of × 1000, the surface appeared mirror-like, flat and scratch-free ", although it is certain that the polish would have induced considerable superficial work damage[55, 56]. Application of the Schell etch[61] produced a set of etch pits which were subsequently shown to be due to grown-in dislocations only. Careful examination of Fig. 1 in this reference[60] shows several indistinct white lines (scratches?) in the background areas, which probably indicates that the Schell etch is relatively insensitive to the work damage. By contrast many defect etches attack work damage to such an extent that it is imperative to remove it completely if a meaningful estimate of grown-in defect content is required.

## 5. ETCH COMPOSITIONS

The section will be considered in three parts: solutions for polishing and general bulk removal, electrolytic and jet etch solutions and solutions specifically for defect structures.

### *5.1. Polishing etches*

These etches are important in the preparation of slices for subsequent processing or assessment. Polished samples are required for epitaxial growth and device fabrication, for infrared microscopy, electron microscopy and X-ray topography and also to provide a featureless surface which is free of work damage and suitable for etching in the defect etches. For many of these applications the polished faces must be flat, and in order to establish and maintain planarity a mechanical/chemical technique is employed. The specimen is mounted on a disc and continuously rotated in contact with a polishing pad impregnated with etch solution. One of the first etch solutions used was a mixture of bromine and methanol; the composition of the etch depended on the laboratory involved and the finish required[62, 63]. Chlorine dissolved in organic solvents[64] has been tried but proved somewhat hazardous (the heat of reaction between chlorine and methanol is sufficient to ignite the methanol under certain circumstances).

Dyment and Rozgonyi[65] have discussed limitations of the bromine–methanol procedure; firstly it is unable to polish the {111} Ga face non-preferentially, and secondly its comparatively rapid ageing characteristics result in non-reproducible polishing rates with time. These authors investigated the syton polishing technique which was developed for silicon[66] but concluded that as gallium arsenide was softer than silicon the syton method did not give a blemish-free surface. Instead they claimed that a polish consisting of 1:700 $NH_4OH$:$H_2O_2$ (ammonia to make the solution neutral) gave a better finish than either syton or bromine–methanol.

Dyment and Rozgonyi[65] briefly discuss the mechanism of chemical/mechanical polishing using the ammonia–peroxide solution and conclude that, since the solution alone merely forms a thin oxide film over the surface, polishing takes place by removal of the oxide layer mechanically. This is followed by oxidation of fresh GaAs in contact with the etch, whereupon the process repeats. Oxidation occurs preferentially at the high spots which are in the most intimate contact with the wet pad. We consider that this explanation is essentially correct, with the additional factor that the oxide particles themselves act as a fine abrasive within the etch solution, forming a polishing slurry.

Sodium hypochlorite diluted with water over a range from 1:15–25 has been successfully used to polish all low-index GaAs orientation substrates[64]. Improvement to the surface finish can be obtained by altering the composition of the etch towards the end of the polishing schedule[67] from 1:12 to 1:40 and by applying bursts of 2.5 ml solution to the pad at 1 min intervals[64].

Bradley[68] has shown that the pressure applied between the sample and the pad is fairly critical for achieving good polishing. Two of his experimental curves for removal rate *versus* pressure are shown in Fig. 5; the best polishing was obtained at the minima of these curves, *i.e.* at the slowest removal rate. As the mechanical/chemical etch does not eliminate work damage a final “free etch” treatment in a chemical polish is necessary for those substrates which are to be used for epitaxy or assessment. For a satisfactory free polish etch a highly viscous solution is desirable[69]. The viscosity determines the in-diffusion of reactants and out-diffusion of reaction products. Sulphuric acid mixtures are relatively viscous and solutions containing other viscous components such as glycerine have been reported. The favoured free etch solution is a mixture of sulphuric acid, hydrogen peroxide and water in ratios 3:1:1 or 5:1:1 used at an elevated temperature (45°–60 °C). Iida and Ito have studied a range of compositions, temperature and stirring rates for this free etch applied to a group of different orientation substrates[70]. They conclude that good polishing can be obtained under correct conditions for all low-index orientations with the exception of the {111} Ga surface (see Section 5.3.2).

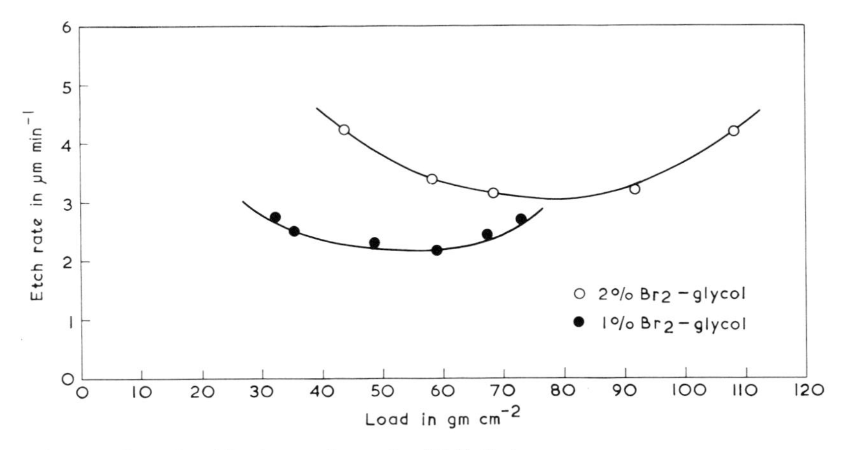


Fig. 5. Effect of pad load on etch rate for {001} GaAs.

Although most workers use a chemical/mechanical polish followed by a free etch, Grabmaier and Watson[71] claimed that a mixture of 1:4 bromine: acetic acid would polish (001) GaAs after lapping. They also reported that 49:11 $H_3PO_4$:$HNO_3$ at 60 °C polished {111} As and {111} Ga surfaces. This latter claim is important, because the foregoing polishes are unsatisfactory for {111} Ga surfaces (most of them etch preferentially).

*5.2. Electrolytic etches and jet etching*

As already stated previously any process which affects the creation of electrons and holes at the sample surface will change the etching rate of a particular solution. Electrolytic etches come into this category as the etching rate can be directly affected by current passing through the solution. Pleskov[72] has measured the electrochemical properties of gallium arsenide in alkali and acid electrolytes and obtained a mirror smooth finish for both p-type and heavily doped n-type GaAs in a 10–40% KOH or NaOH solution at a current density of 1–5 A $cm^{-2}$. Harvey[73] found no substantial difference in the electrochemical behaviour of {111} Ga and {111} As surfaces in KOH and $HClO_4$ solutions. Haisty[74] has taken the process a step further by studying the effect of light on samples in dilute electrolytes. He used the method to reveal etch pits on {111} Ga faces. Iiyama *et al.*[75] have described a rotating disc apparatus which has given a smooth polish finish to {111} Ga surfaces with a variety of electrolytes. This is also one of the advantages claimed by Unvala *et al.*[76] for the technique of "jet" etching. This method of etching is normally applied to the shaping of GaAs samples for electron microscopy specimens and can be performed either as a jet chemical polish or as a jet electrolytic polish. Bicknell[77] has used a rotating jet-thinning apparatus to thin and polish GaAs samples. A continuously chlorinated methanol solution was prepared at the jet orifice itself; examination showed that the etching was relatively insensitive to local substrate variations and so produced single curvature hemispherical depressions. The advantage of these specimens is that they may be subsequently treated with solutions which differentially attack p and n regions and may be examined by both optical and transmission electron microscopy.

### *5.3. Defect-revealing solutions*

#### *5.3.1. Etch calibration*

The objective of defect etching is the production of characteristic features at the specimen surface (*e.g.* etch pits) which can be correlated with specific defects present in the surface region (*e.g.* dislocations). Establishment of the correlation is known as *calibration* of the etch, and it is essential if the etch is required for quantitative assessment purposes. Various direct methods have been devised for calibration of dislocation etches; methods for correlating other defects with etch features have usually been indirect and inferential.

Nye[78] showed that the equilibrium number $N$ of dislocations of one sign present in a uniformly plane bent crystal was simply related to the radius of curvature $\rho$ and the Burgers vector of the dislocation $\boldsymbol{b}$ as

$$N = 1/\rho b \cos\theta$$

where $\theta$ is the angle between the Burgers vector and the neutral plane. For this equation to apply, slip must occur on one slip system only. Abrahams and Ekstrom[60] prepared rectangular single-crystal bars of GaAs oriented so that, firstly, plastic deformation at 700 °C under symmetrical four-point loading produced essentially single glide on the $(1\bar{1}1)$ $[\bar{1}01]$ system and, secondly, two opposite faces of the bars were $\{111\}$ surfaces containing traces of the active $(1\bar{1}1)$ slip plane. By bending the specimens to a series of different curvatures and then etching the $\{111\}$ surfaces they were able to correlate the calculated dislocation densities with observed densities of etch pits, for the particular etch used. By this means, Abrahams and Ekstrom[60] calibrated the Schell etch[61] as a dislocation etch for $\{111\}$ Ga surfaces. A later investigation of another etchant (the A/B etch) by Abrahams and Buiocchi[30] used an interesting modification of this technique in which two crystal bars in the same orientation were bent to the same radius of curvature, but one was bent convex upwards and the other convex downwards (the method of reverse bending[29]). The majority sign dislocations accommodating the bending will be of $\alpha$ type for one curvature and $\beta$ type for the other (*cf.* Section 4.3.1). Abrahams and Buiocchi were able to demonstrate that the A/B etch revealed only the (As) $\beta$ type of dislocation. Subsequent annealing experiments on the bent samples caused copper decoration of both types of dislocations, and both types could then be etched.

It is possible to use one etch, which has been calibrated, as a standard for establishing the calibration of other etches. This method was used by Abrahams and Ekstrom[60] to calibrate a $\{111\}$ Ga etch devised by White and Roth[17]. The Schell etch was used to establish a set of etch pits at the $\{111\}$ Ga surface. The sample was then lightly polished and etched in the White and Roth solution. Comparison of identical areas showed that new pits formed at the sites of the Schell etch pits, and hence this established the validity of the White and Roth solution as a dislocation etch. Abrahams was also able to utilise the calibration of pits on a $\{111\}$ Ga face[60] as a standard for establishing the calibration of an etch which produced pits on the opposite {111} As face[79]. His method assumed that patterns of dislocations emerging at one face would be similar to the patterns on the opposite face, provided the sample was very thin.

*5.3.2. Specific defect etches*

Many of the etchants devised for revealing defects in GaAs contain an oxidising agent. It is therefore convenient to group the defect etches under the headings of the particular oxidising reagents present in the solutions. Unusual aspects of the etching behaviour of some of the etchants will be considered in detail in Section 6.

*5.3.2.1. Nitric acid.* Concentrated nitric acid will attack GaAs low-index surfaces, but it is unsuitable as a defect etchant because crystals of $As_2O_3$ precipitate over the surfaces[6]. Dilute nitric acid does not suffer from this disadvantage because the $As_2O_3$ remains in solution. Chronologically, the first GaAs etch using dilute nitric acid was described by Schell[61]. It consisted of 1:2 $HNO_3$:$H_2O$. This is the "correct" Schell etch for {111} Ga surfaces, and was the composition calibrated by Abrahams and Ekstrom[60] (see Section 5.3.1). No etch times or temperatures were given by Schell, although we have found that 1–3 min at 60 °C is sufficient for the formation of reproducible etch pits on {111} Ga surfaces. Other authors have used different dilutions and conditions: Grabmaier and Watson[71, 80] and Grabmaier and Grabmaier[81] used 1:$1\frac{1}{2}$ $HNO_3$:$H_2O$ at 60 °C for 10–20 s, and Richards and Crocker[82] used 1:3 $HNO_3$:$H_2O$. Both of these solutions were quoted as "Schell etch". Our own experiments suggest that alterations to the $HNO_3$:$H_2O$ ratio do not affect the characteristic shapes of the {111} Ga etch pits, which consist of terraced triangular pyramids with sides parallel to <110> substrate directions (see Fig. 7(b)), but that they do alter the etching times required to produce particular-sized etch pits at a constant temperature. Steinemann and Zimmerli[35], for example, found that a 1:$1\frac{1}{2}$ $HNO_3$:$H_2O$ composition used at 60 °C for about 10 s produced similar-sized etch pits (~50 μm) to those formed after 1–8 min at 60 °C in the "correct" (1:2) composition.

Addition of hydrochloric acid alters the pit shapes, as shown by White and Roth[17]. Their etchant consisted of 2:1:2 HCl:$HNO_3$:$H_2O$, which produced circular etch features on {111} Ga surfaces, at sites shown to correspond with the Schell etch pit positions by Abrahams and Ekstrom[60].

Richards and Crocker[82] suggested that acidic oxidising agents were too reactive to develop well-formed pits on {111} As surfaces. They found that an effective etch for both {111} Ga and {111} As surfaces could be obtained by using Ag ions (from $AgNO_3$ solution) to inhibit the rapid attack of an $HNO_3$:HF mixture. Abrahams[79] calibrated a modified version of this etch, which he named the RC-1 etch, by comparison with Schell etch results. The composition of RC-1 was as follows: $2.4 \times 10^{-3}$ molar solution of $AgNO_3$ in 3:2:5 $HNO_3$:HF:$H_2O$. This etch produced triangular pits on {111} As surfaces and circular pits on {111} Ga surfaces.

The cell structure shown in Fig. 6 was delineated by etching the (001) surface of a Cr-doped GaAs crystal in 3:1:6 $HNO_3$:HF:$H_2O$. A slightly different composition, 3:1:4 $HNO_3$:HF:$H_2O$, was used by Plaskett and Parsons[83] to reveal growth striae in Sn-doped GaAs. LeMay[53] found that the Schell etch could

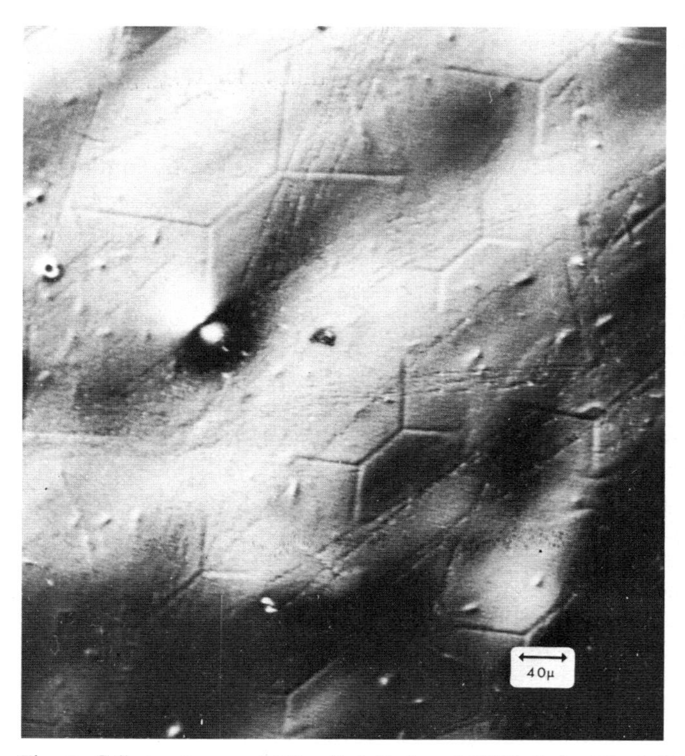


Fig. 6. Cell structure on LEC pulled Cr-doped {001} GaAs. The "ridge" line of the structure lies in <110>; sample etched in 3:1:6 $HNO_3$:HF:$H_2O$. (Copyright © Controller H.M.S.O., London, 1975.)

be used to reveal growth striae on {110} surfaces of Te-doped GaAs, but only after prolonged etching.

*5.3.2.2. Sulphuric acid.* Most of the etches based on sulphuric acid have already been considered in Section 5.1. The work of Iida and Ito[70] dealing with a range of $H_2SO_4$:$H_2O_2$:$H_2O$ compositions showed that, although large regions of the compositional "phase diagram" gave good polish finishes, compositions with high $H_2O$ proportions produced cloudy surfaces. They suggested that this cloudiness was due to the formation of small pits which may have been influenced by crystalline defects in the substrates. We have found that a 3:1:1 $H_2SO_4$:$H_2O_2$:$H_2O$ composition will produce etch pits on the {111} Ga face (see Fig. 7(a) and (b) and Section 6.2 for further details). The 3:1:1 etch composition is normally used as a free etch polish, especially for (001) surfaces (see Section 5.1). However, Bolger[84] has shown by careful optical examination that although the (001) surfaces are essentially flat and polished some defect features are also present. These observations are dealt with further in Section 6.5, but one explanation may involve the effects of illumination of the sample surface on the etching action of the solution. Kuhn-Kuhnenfeld[8] has found that intense local illumination, coupled with the electrical characteristics of the GaAs sub-

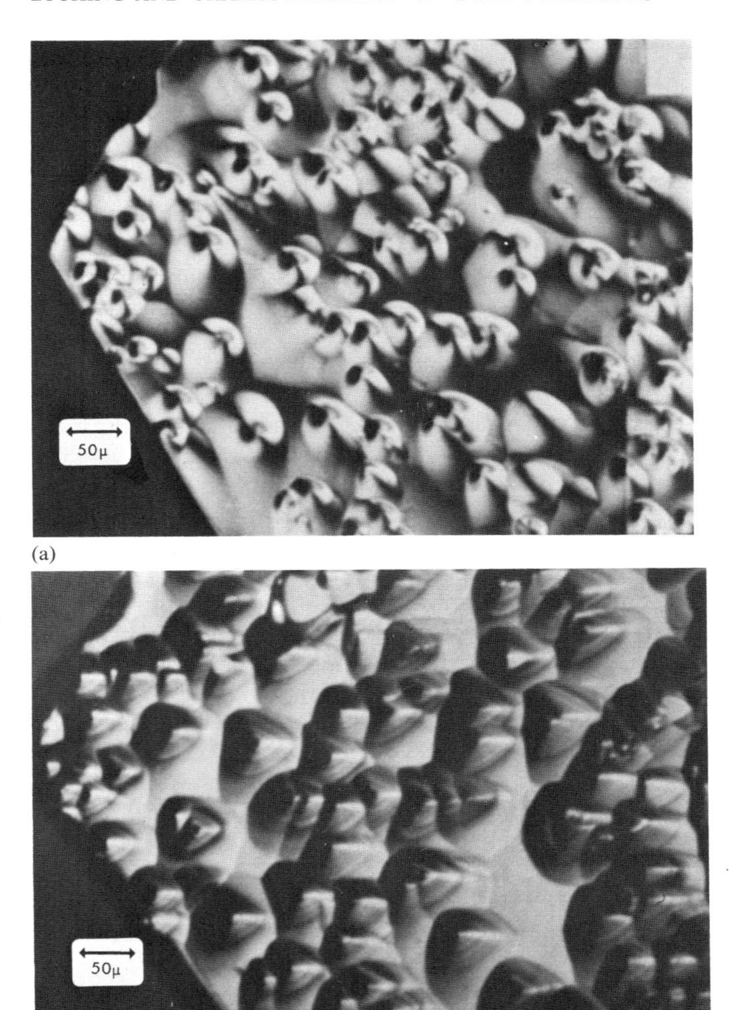


Fig. 7. (a) Etch pits on {111} Ga produced by 3:1:1 $H_2SO_4$:$H_2O_2$:$H_2O$. (b) Identical area to (a) after treatment in Schell etch (2 min at 60 °C).

strates, can entirely alter the way in which 3:1:1 $H_2SO_4$:$H_2O_2$:$H_2O$ attacks (001) surfaces. He was able to reveal dislocations and striations with excellent resolution, even in undoped or semi-insulating samples. Measurements of photovoltages as a function of the intensity of illumination showed that surface photovoltaic effects could be responsible for changes in etch reactions at the illuminated areas.

*5.3.2.3. Chromium trioxide.* The most widely used defect etch for (001) GaAs surfaces is that due to Abrahams and Buiocchi[30], and commonly referred to as the A/B etch. Its composition is 8 mg $AgNO_3$, 2 ml $H_2O$, 1 g $CrO_3$, 1 ml HF.

This etch can be used to etch all low-index crystallographic faces, and Abrahams and Buiocchi calibrated it for {111}, {110} and {001} GaAs surfaces by a variety of different techniques[30]. Both α(Ga) and β(As) dislocations could be revealed at {111} surfaces by etch pits, provided that the dislocations had been suitably decorated (*e.g.* during growth of the crystal). In the absence of impurity decoration, only "clean" β(As) dislocations were etched. The A/B etch can also reveal striations and stacking faults on (001) surfaces (a modification was required to delineate stacking faults on {111} As surfaces). We have found that s-pits (Section 4.3.2) are formed by the A/B etchant in Te-, Se- and Si-doped GaAs, and that it can produce characteristic etch pits ("boat pits") at work-damaged areas of (001) substrates (see Section 6.3). In view of the almost universal use of (001) orientation substrates for device processing, an etch such as the A/B solution assumes great importance, and it is essential to understand the features revealed by it at (001) surfaces. Etch pits do *not* appear at the sites of emergent dislocations, unless the latter happen to be exactly in [001] directions (as is the case for some dislocations in (001) epitaxial layers[30]). These, and other unique features of the A/B etch, will be described in Section 6.

The Sirtl etch[85], well known as a defect etch for silicon, has been used for GaAs. It consists of a mixture of chromic acid and hydrofluoric acid and was found by Meieran[57] to delineate striations on {111} Ga and {111} As faces after a 2 min application. More recently Olsen *et al.*[33] found that the Sirtl etch produced rectangular hillocks on (001) and $(00\bar{1})$ GaAs surfaces. They argued that the directions given by the unequal sides of the hillocks, which lay parallel to [110] and $[1\bar{1}0]$ directions, could be used to distinguish between these two directions in the (001) plane. Although the Sirtl etch figures for (001) Si (which are square based) have been related to dislocations, no such calibration has yet been made for the GaAs rectangular hillocks.

Cardwell and Stirland[86, 87] developed an etchant for (001) GaAs surfaces which consisted of 3:3:1 $CrO_3$:HF:$HNO_3$. X-ray transmission topography in conjunction with optical observation of etched surfaces showed that this etchant produced approximately circular etch hillocks at emergent dislocations.

*5.3.2.4. Potassium hydroxide.* The etch described by Grabmaier and Watson[71] does not fall into the category of etch solutions. It is a single component etchant consisting of molten KOH maintained at 300 °C. Immersion for 2–3 min was sufficient to develop etch pits on {111} Ga, {111} As and {001} surfaces. Grabmaier and Watson calibrated the etch for these three surfaces. We have found that their results can be reproduced for (001) surfaces provided that a pre-etch polish had been given in bromine–methanol or bromine–acetic acid mixtures. Other pre-etch polishes such as 3:1:1 $H_2SO_4$:$H_2O_2$:$H_2O$ consistently inhibited the action of the molten KOH.

## 6. SPECIAL ASPECTS OF THE ACTIONS OF ETCHANTS

This section considers experimental and interpretational peculiarities which can be encountered while following some of the etching procedures discussed in earlier sections.

*6.1. Surface contamination*

The precipitation of crystallites of $As_2O_3$ on GaAs surfaces during reaction with strong nitric acid has already been mentioned. This is gross surface contamination, readily detected and remedied. In this section we draw attention to precipitation on a molecular scale, the sources of the contamination being (1) doping elements present in the GaAs crystal, and/or (2) components of the etching solution.

Hvålgard *et al.*[88] reported the gradual accumulation of Te dopant on substrate surfaces as the substrate was progressively etched in a variety of different etchants. They measured Te contamination levels 5–10 times higher than in the bulk material. A similar result was obtained with Sn-doped GaAs, except when HF-based etches were used. In these solutions the dopant was more readily removed because Sn forms strong complexes with fluorine.

Larabee[89] studied the absorption of a number of metal ions on GaAs surfaces during nitric acid washes. The metal ions, present in the etching solutions in fractional ppm quantities only, readily formed monolayers on the substrate surfaces. Larabee gave data on the electrochemical deposition of Cu, Ag, Hg, Au and Zn, and he showed that in some cases the metal ions were irreversibly adsorbed.

Surface contamination from etch components has been found by Newton[90] for the reaction of GaAs with 3:1:1 $H_2SO_4$:$H_2O_2$:$H_2O$. Using radiotracer methods involving $^{35}S$ he showed that sulphur atoms could be determined on GaAs substrate surfaces at typical levels of $2 \times 10^{-8}$ g, after a standard (15 min) etch in the 3:1:1 polishing solution, followed by copious water washing.

The effects described here may need to be considered when epitaxial layers are to be grown on etched substrates to stringent specifications.

*6.2. Surface treatment before etching*

The necessity for removing superficial work damage before employing defect etchants to reveal the intrinsic defect content of a sample has already been stressed. Usually, work damage will be removed by a pre-defect etch treatment, which essentially will be a polish etch (free etch). Unfortunately, there is evidence that some of the pre-etch treatments affect the performance of the subsequent defect etchants. We have already mentioned (Section 5.3.2.4) that a molten KOH etch did not operate successfully on (001) surfaces which had been pre-treated in 3:1:1 $H_2SO_4$:$H_2O_2$:$H_2O$ or in solutions containing $HNO_3$. Etch pits were not formed, and the sample surface became badly stained. If the pre-treatment solution consisted of bromine–methanol or bromine–acetic acid[71], the desired etch pits appeared reproducibly.

We have also found that the Schell[61] etch will not attack {111} Ga surfaces which have been given a chemico-mechanical pad-polish using 1:700 $NH_4OH$:$H_2O_2$. However, the Schell etch will produce "orthodox" etch pits on these surfaces if they are treated prior to its use with 3:1:1 $H_2SO_4$:$H_2O_2$:$H_2O$. Examination of the {111} Ga surfaces after treatment with the latter mixture showed that it had produced circular etch pits; subsequent treatment in the Schell etchant converted the circular etch pits into triangular etch pits. Figure

7(a) and (b) shows the same area of a {111} Ga surface after attack firstly by the 3:1:1 $H_2SO_4$:$H_2O_2$:$H_2O$ etch and secondly by the Schell etch. The micrographs demonstrate that there is a one-to-one correspondence between circular and triangular pits. Since the Schell etch is a calibrated etch (see Section 5.3.1) these observations serve to calibrate 3:1:1 $H_2SO_4$:$H_2O_2$:$H_2O$ as a dislocation etch for {111} Ga surfaces of gallium arsenide.

It is clear from the foregoing that the action of some defect etches is critically dependent on the nature of the surfaces produced by the pre-etch treatment(s). Schwartz[91] has shown that various surface films can be formed on GaAs surfaces by reaction over long periods with methanol, water, hydrogen peroxide, sulphuric acid and nitric acid. $Ga_2O_3.H_2O$ was detected by X-ray analysis after (001) surfaces were reacted with $H_2O$ at 60 °C for 6 days, whereas {111} As surfaces were completely converted to $As_2O_3$ powder after storage in $HNO_3$ for 5 days. Reaction with $H_2SO_4$ for 6 days produced $Ga_2(SO_4)_3$ at (001) surfaces. The various samples examined by Schwartz were prepared initially using a $Br_2$–methanol polish treatment[63], and this always resulted in a residual (oxide) layer at the specimen surfaces of 25–50 Å thickness. Hasegawa *et al.*[92] have recently reported that anodic oxide layers formed in tartaric or citric acid mixed with glycol are insoluble in concentrated nitric acid and halogen–alcohol mixtures.

*6.3. Action of the A/B etchant at (001) GaAs surfaces*

Abrahams and Buiocchi[30] examined the (001) surfaces of epitaxial GaAs layers grown on (001) substrates after etching in the A/B etchant, and they found that conical etch pits of approximately 10 μm diameter were formed. These conical pits did not occur when bulk (001) GaAs substrate material was etched in the A/B etchant. In this case the A/B etchant produced lines (largely parallel to <110> directions) or etching "tails". The lines were assumed to be dislocations nearly parallel to the surfaces. The tails consisted of strings of overlapping pits of increasing diameters and were believed to be due to dislocations inclined to the surfaces, impurity segregation having caused enhanced local attack and pit formation along the dislocations. Later work by Huber and Champier[93] showed that there was some agreement between line structures produced on (001) surfaces by the A/B etchant and dislocations seen on X-ray topographs of identical areas. Stirland and Ogden[94] analysed some of the processes which occur during the development of the A/B etching structures, and the results of more recent observations[95] have confirmed this analysis. Stereo-transmission X-ray topography[96] was used to establish the volumetric dislocation distribution within a selected region of the original (unetched) substrate. One surface of the sample was then etched in the A/B etchant, and micrographs from the selected region were obtained. Further X-ray topographs were taken, and further etching was performed. At each etching stage a corresponding topograph of the sample was taken. The results of these experiments showed that the action of the A/B etchant was to produce ridges about 1–5 μm high at dislocation lines (the ridges were seen as black/white lines on interference contrast micrographs). The most surprising observation was that these ridges were preserved without obvious alteration during the course of subsequent etching, even when the A/B etchant had removed a thickness of approximately 40 μm more material from the surface. Stirland and Ogden[94] called this effect a dislocation "etch memory", because the ridges represented dislocation lines which no longer existed. The optical micrographs represented a two-dimensional projection (on the surface attained after etching) of dislocations which had been present in the volume of material removed by the etch. This was demonstrated by superimposing each transmission X-ray topograph of the material remaining after etching and the corresponding micrograph of the etched surface; the combination of lines and ridges was identical with the dislocation lines on the transmission topograph of the original unetched sample. Effectively, the optical micrographs of A/B etched surfaces were equivalent to transmission X-ray topographs of the volume of material which had been removed by the etch. They could thus be used to determine dislocation densities. The procedure required the measurement of the thickness of material removed during the etching; use of a Talysurf surface probe[97] at steps produced by waxed-off regions provided a rapid and accurate method for this determination.

The number of intersections per unit length made by the ridges was counted from micrographs; knowing the depths removed by etching it was possible to calculate dislocation densities. These values represented the number of dislocations intersecting unit area *perpendicular* to the (001) specimen surfaces. "Orthodox" etching behaviour as typified by the Schell etch reaction at {111} Ga surfaces (and apparently by the A/B etch attack on (001) epitaxial layers[30] in which dislocations intersect the (001) surface exactly at 90°) produced etch pits at dislocations intersecting the surfaces, and thus provided dislocation density values which represented the number of dislocations threading unit area of the surfaces.

Figure 8 shows a typical area of an Si-doped (001) GaAs substrate after

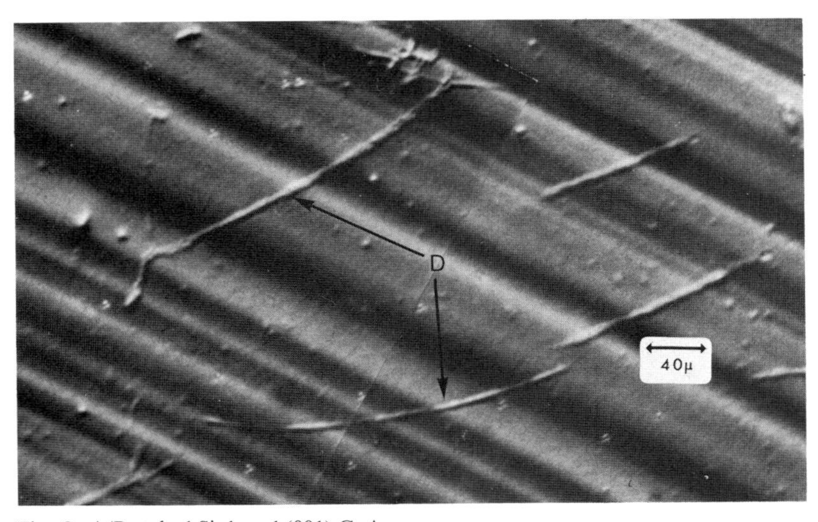


Fig. 8. A/B etched Si-doped (001) GaAs.

5 min treatment at room temperature in the A/B etchant, which removed about 12 μm from the original surface. The narrow black and white lines marked D are delineating the (projected) dislocation arrangement as it was originally present in the 12 μm of material removed by etching. The broad diffuse dark and light bands are growth striations, due to segregation of the added impurity. Figure 9(a) and (b) demonstrates the preservation of features ($E_1$ and $E_2$) by the etch memory process for a Cr-doped (001) GaAs substrate. A first etch which removed 10 μm (Fig. 9(a)) was followed by removal of a further 15 μm (Fig. 9(b)) by a second etch of the identical area. The circular features (*e.g.* $P_1$ and $P_2$) are pits, often situated in the dislocation ridges, and are believed to arise from impurities segregated at the dislocations during growth. More

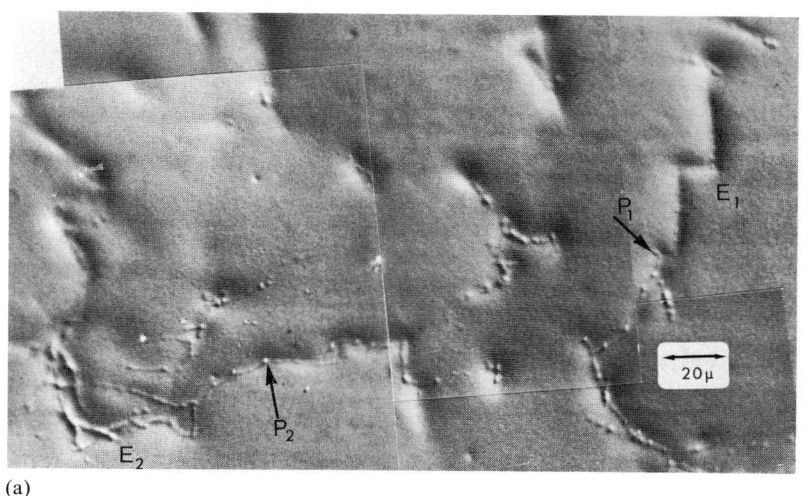


(a)

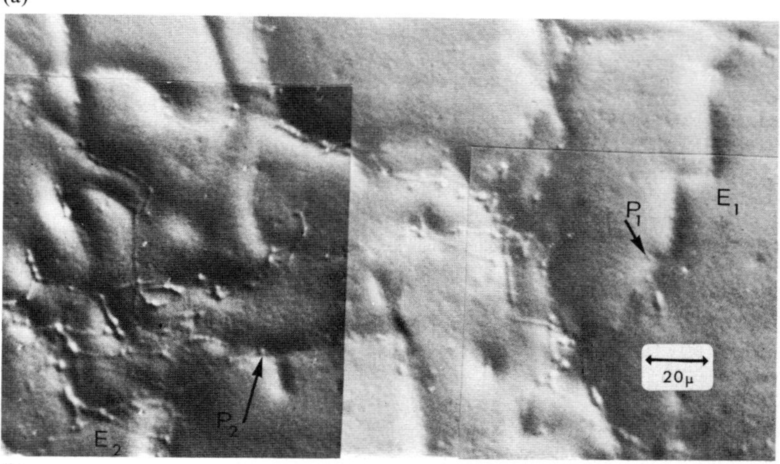


(b)

Fig. 9. (a) A/B etched Cr-doped (001) GaAs, 10 μm removed by etch. (b) Identical area to (a), 25 μm removed by etch.

pronounced examples of decoration of dislocations by impurity segregation have been found with group VI dopants such as Se and Te. Figure 10 shows a region of an Se-doped (001) GaAs substrate (carrier concentration $1.5 \times 10^{18}\ \text{cm}^{-3}$) in which the dislocation fragments were heavily decorated, as indicated by the enhanced pitting of the dislocation lines (contrast with the Si-doped sample of Fig. 8). In this sample pitting also occurred in dislocation-free regions, but examination showed that a pit-free region of approximately 50–100 μm surrounded each dislocation. Figure 11 shows a single dislocation in a Te-doped (001) GaAs substrate (carrier concentration $2.4 \times 10^{18}\ \text{cm}^{-3}$), at higher magnification. It can be seen that the dislocation was heavily pitted along its length and that the size and shape of the pits were different at ends $D_1$ and $D_2$. Experiments on repeated etching of identical areas have shown that the shallow elongated shapes at end $D_1$ and at positions $P_2$ evolved after several minutes etching from smaller, nearly circular, etch pits which appeared at the start of etching. Typical examples are shown at end $D_2$ and at positions $P_1$. Thus the part of the dislocation at $D_2$ had just been reached when etching was terminated, whereas the part at $D_1$ intersected the initial surface and had been attacked from the beginning of the etching treatment.

The pits at $D_1$ and $P_2$ are elongated along one <110> direction only; examination of similar samples etched on both {001} faces showed that the <110> elongation direction of the pits on one face was orthogonal to the <110> elongation direction of the pits on the opposite face. We believe that these pits are produced when the A/B etch attacks defects which are probably impurity precipitates formed primarily by the dopant element and are not associated with dislocations, for reasons similar to those proposed by Iizuka[44] for the occurrence of s-pits in GaP. The key difference between the s-pits shown

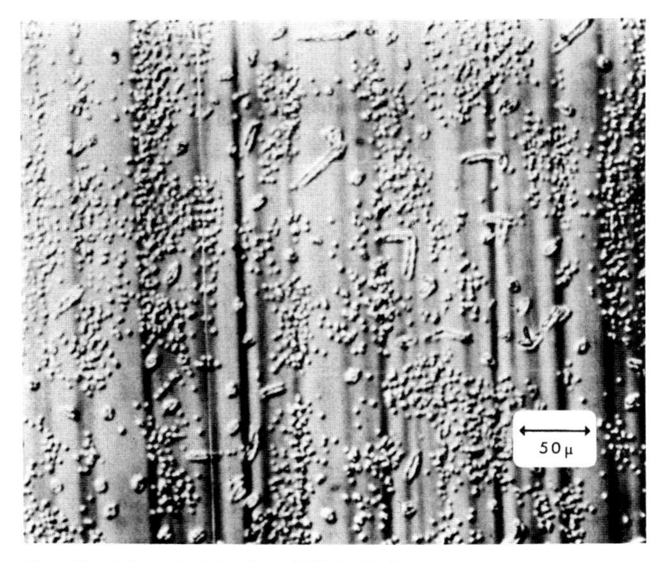


Fig. 10. A/B etched Se-doped (001) GaAs.

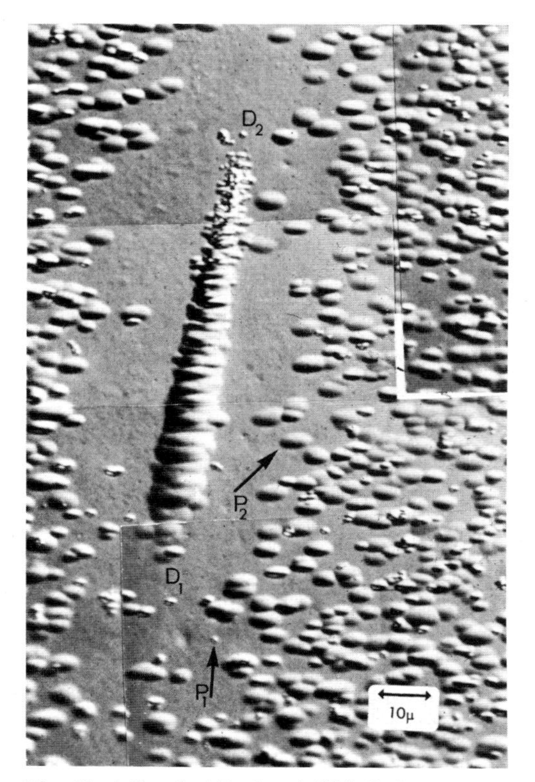


Fig. 11. A/B etched Te-doped (001) GaAs.

by Iizuka and his coworkers[37, 38, 44] for GaP and the pits shown in Fig. 11 is that, whereas the former appear approximately circular, the latter are evidently elongated. However, all the GaP s-pit micrographs were obtained from {111} surfaces, whereas the GaAs s-pits were from (001) surfaces. We consider that the asymmetry shown by the (001) surface pits is related to differences between the A/B etch rates at {111} Ga and {111} As surfaces, as reported by Abrahams and Buiocchi[30]. As soon as pitting by penetration of the etchant into the volume of an (001) surface specimen occurs (*e.g.* at defects) then the development of asymmetric pit shapes is inevitable if differences exist between the attack rates at symmetrically disposed, but non-equivalent, inclined low-index planes such as the {111} Ga and the {111} As (see Section 6.4).

A gross form of etch pit asymmetry can be seen by the attack of the A/B etchant at work-damaged (001) surfaces. Figure 12 shows an A/B etched (001) GaAs substrate after a mechanico-chemical lapping treatment. Surface scratches have been delineated as close-packed or separated lines of pits, all elongated along one specific <110> direction. At high magnifications it can be seen that some of the elongated pits have sharp-edged bottoms ("boat pits"), as in Fig. 13. The presence of "boat pits" is an indication that work damage is present at a surface, but the specific defects which develop into boat pits are

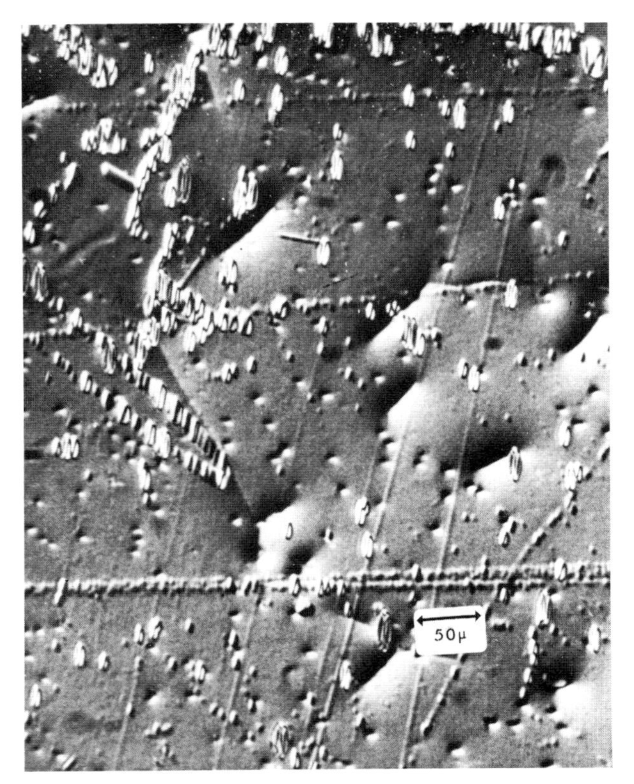


Fig. 12. A/B etched Si-doped (001) GaAs after lapping treatment.

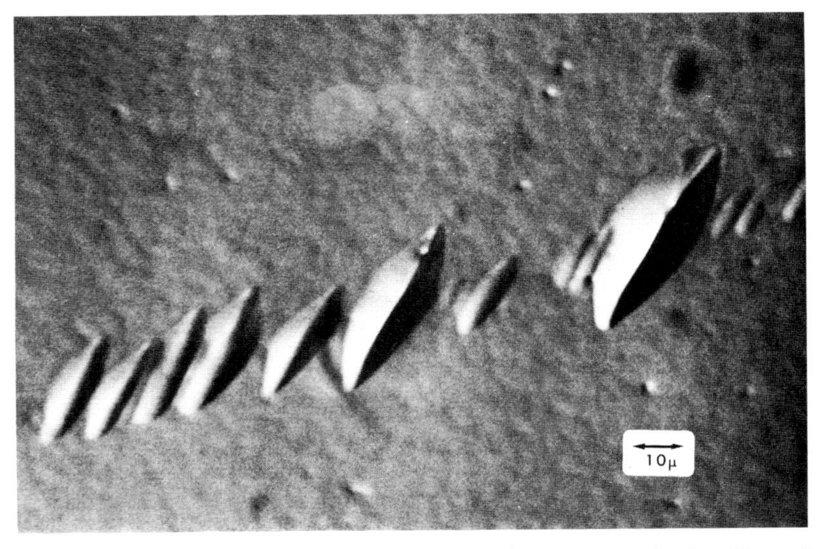


Fig. 13. A/B etched Si-doped (001) GaAs after lapping treatment showing "boat pits" with long axis in <110>.

not known. Huber and Champier[93] suggested that elongated pits, which they named long microcavities, were due to etching of either dislocation loops or stacking faults.

Finally, it should be mentioned that the A/B etchant behaves in an "orthodox etch" manner at {111} Ga and {111} As surfaces, by producing etch pits which correspond to the positions of emergent dislocations[30]. However, the behaviour at {110} surfaces appears to be similar to that at (001) surfaces.

### *6.4. Comparison of A/B and molten KOH etches at (001) GaAs surfaces*

Use of the A/B etch can provide a lot of information about the defect structure of GaAs. There are, however, certain ambiguities in the interpretation of etch features on (001) surfaces because some of the complex dislocation patterns are due to the "memory" effect[94] of the etch. Molten KOH does not exhibit a memory effect but it is a less attractive etch to use because the reaction is at 300 °C. We have made a comparison of the etch patterns on (001) slices of GaAs using an A/B etch followed by molten KOH on the same sample area. In this experiment the A/B etch pits were removed by a polish in bromine–methanol before the specimen was etched in the molten KOH. On a second sample we performed the etchings in the reverse order. The aim was to determine if the etch pits produced by molten KOH bore any relation to the lines produced by the A/B etch.

The results of the first experiment (A/B followed by molten KOH) are shown in Fig. 14(a) and (b). There is almost a one-to-one relation between pits and lines. The lines do not appear to be dislocation half-loops with both ends of the loop emerging at the same surface, since etch pits occur at only one end of the line. The boat-shaped pits produced by the A/B etch did not reappear during the KOH etching.

When the experiment was performed in the reverse order the etch pits produced by an initial KOH etch (Fig. 15(a)) were not polished out before using the A/B etch (Fig. 15(b)). The results from this experiment demonstrate that the <110> elongated pits discussed in the previous section can result from the attack of the A/B etchant at existing pits in an (001) surface; the presence of defects at the sites of the existing pits is neither sufficient nor necessary *per se* to generate the elongated pits, although defects are required to generate the KOH etch pits.

### *6.5. Defect etch action of polishing solutions*

In Section 6.2 we presented evidence to show that 3:1:1 $H_2SO_4$:$H_2O_2$:$H_2O$ etchant acted as a dislocation etch for {111} Ga surfaces, although it is normally employed as a polish etch for (001) surfaces. Bolger[84] has found that this solution can also reveal dislocations and s-pits on (001) GaAs surfaces under conditions which are not well understood at present, but are probably related to the illumination experiments of Kuhn-Kuhnenfeld[8]. Figure 16(a) and (b) shows (001) GaAs substrate surfaces after a "polish etch" treatment in the 3:1:1 $H_2SO_4$:$H_2O_2$:$H_2O$ mixture, followed by an A/B etch. Adjustment of the interference contrast[98] separated the contrast of the A/B etch features (Fig. 16(b))

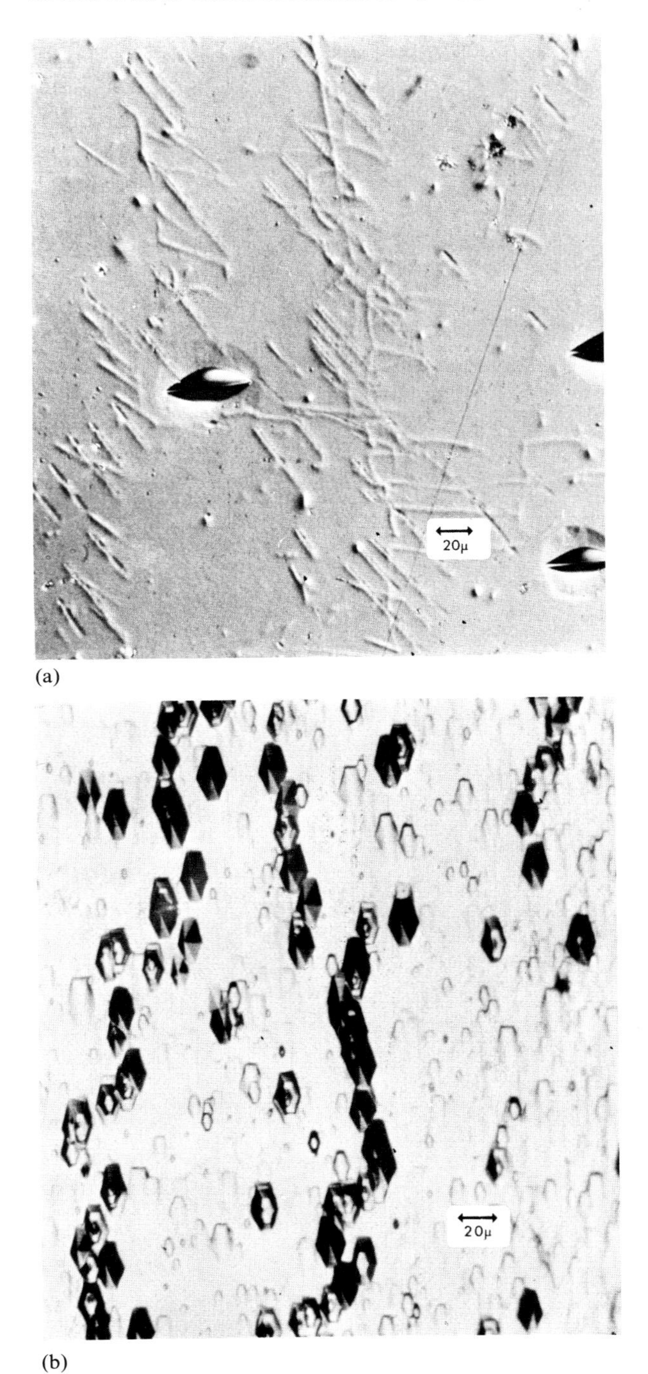


Fig. 14. (a) A/B etched Si-doped (001) GaAs. (b) Identical area to (a) etched in molten KOH after removal of A/B etch features. (Copyright © Controller H.M.S.O., London, 1975.)

from those produced by the "polish etch", indicating that the two forms of attack were dissimilar (grooves on the "polish etched" surface became ridges on the A/B etched surface).

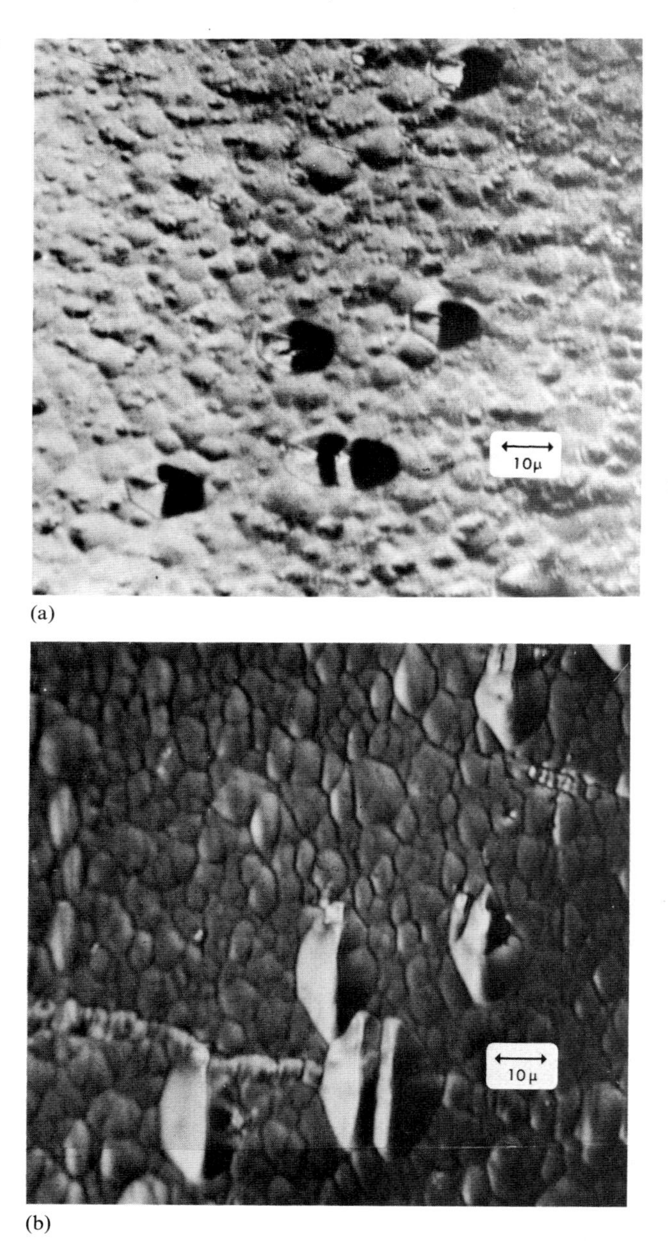


(a)

(b)

Fig. 15. (a) Molten KOH etched Si-doped (001) GaAs. (b) Identical area to (a) etched in A/B solution without removal of molten KOH etch features.

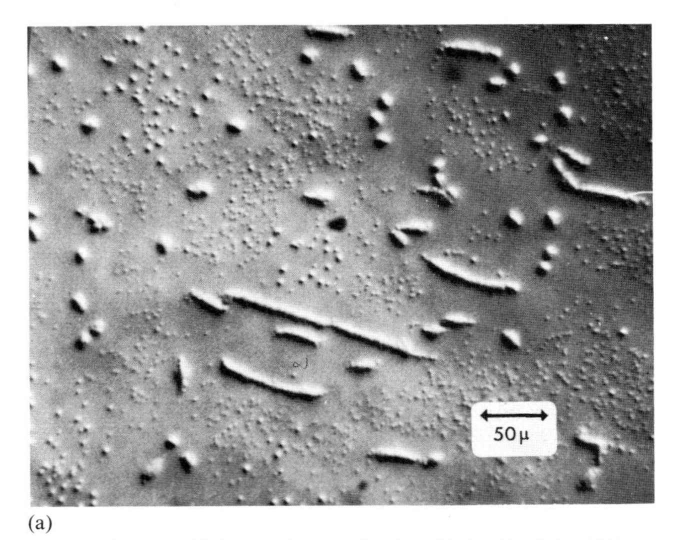


(a)

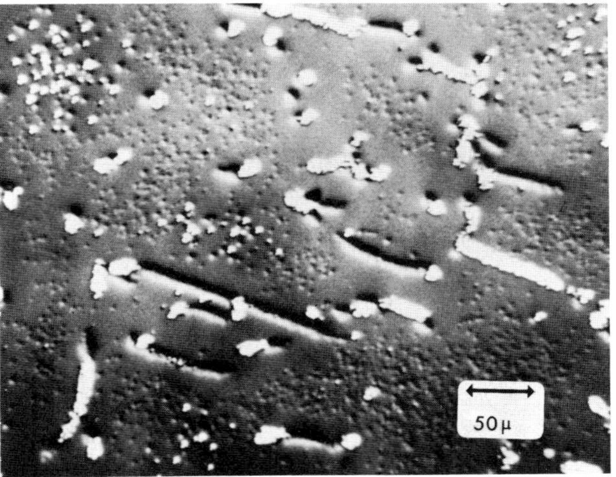


(b)

Fig. 16. (a) (001) GaAs after "polish" etch with 3:1:1 $H_2SO_4$:$H_2O_2$:$H_2O$. (b) Identical area to (a) after A/B etch.

## 7. CONCLUSIONS

In this review we have attempted to cover those aspects of GaAs etching procedures and reactions which are important to the material scientist and which may have consequences for the device manufacturer. There are still some problem areas. For example, there are no precise data on vacancy concentrations or vacancy–impurity complexes present in substrate material which can be related to etching effects. There are large discrepancies in estimates of the depths of work damage, and even difficulties in the formulation of what constitutes work damage. Some of the details involved in impurity precipitation, such as the extent of the denuded regions around dislocations, are not well understood.

In practical terms there have been difficulties in reproducing particular etching effects by the straightforward application of reported procedures. We now believe that one important factor, which does not appear to have been considered hitherto, is the pre-etch treatment of the specimen surface. The behaviour of the molten KOH etchant at surfaces subjected to different pre-etch treatments is a good example. We have also shown that some specimens can be selectively attacked at defects by solutions which usually act as free etch polishes.

However, in spite of these uncertainties, it is evident from the large amount of information spread throughout the literature of the past two decades that the defect content of GaAs single crystals, although complicated, has been well characterised. To some extent this situation has been helped by other physical examination methods such as X-ray topography and transmission electron microscopy, but it is clear that the etch method, in combination with optical microscopy, can yield valuable information on its own.

ACKNOWLEDGMENTS

We are pleased to acknowledge the help of various colleagues who have allowed us to report their unpublished results, in particular D. Bolger, R. R. Bradley and J. Knutton. Special thanks are due to J. B. Mullin for the initiation of some of the etching studies, and for many valuable discussions. We acknowledge the Plessey Company Limited for permission to publish, and Procurement Executive, Ministry of Defence, for support of parts of the work.

REFERENCES

1 J. W. Faust, Jr., in H. C. Gatos (ed.), *The Surface Chemistry of Metals and Semiconductors*, Wiley, New York, 1966, p. 151.
2 T. M. Buck, in H. C. Gatos (ed.), *The Surface Chemistry of Metals and Semiconductors*, Wiley, New York, 1966, p. 107.
3 B. W. Batterman, *J. Appl. Phys., 28* (1957) 1236.
4 H. C. Gatos and M. C. Lavine, *J. Phys. Chem. Solids, 14* (1960) 169.
5 T. H. Yeh and A. E. Blakeslee, *J. Electrochem. Soc., 110* (1963) 1018.
6 D. F. Kyser and M. F. Millea, *J. Electrochem. Soc., 111* (1964) 1102.
7 D. R. Turner, *J. Electrochem. Soc., 107* (1960) 810.
8 F. Kuhn-Kuhnenfeld, *J. Electrochem. Soc., 119* (1972) 1063.
9 R. W. Haisty, *J. Electrochem. Soc., 108* (1961) 790.
10 J. B. Mullin, A. Royle and B. W. Straughan, *Proc. 3rd Int. Symp. on Gallium Arsenide and Related Compounds, Aachen, 1970*, Inst. Phys., London, 1971, p. 41.
11 B. W. Straughan, unpublished results, 1973.
12 B. Tuck, *J. Mater. Sci., 10* (1975) 321.
13 D. B. Holt, *J. Phys. Chem. Solids, 23* (1962) 1353.
14 H. C. Gatos and M. C. Lavine, *J. Electrochem. Soc., 107* (1960) 427.
15 C. M. H. Driscoll, A. F. W. Willoughby, J. B. Mullin and B. W. Straughan, *5th Int. Symp. on Gallium Arsenide and Related Compounds, Deauville, 1974*, Inst. Phys., London, 1975, p. 275.
16 D. B. Holt, *J. Phys. Chem. Solids, 27* (1966) 1053.
17 J. G. White and W. C. Roth, *J. Appl. Phys., 30* (1959) 946.

18 K. F. Hulme and J. B. Mullin, *Solid-State Electron.*, *5* (1960) 211.
19 J. F. Dewald, *J. Electrochem. Soc.*, *104* (1957) 244.
20 M. E. Straumanis and C. D. Kim, *Acta Crystallogr.*, *19* (1965) 256.
21 D. B. Holt, *J. Mater. Sci.*, *1* (1966) 280.
22 H. R. Potts and G. L. Pearson, *J. Appl. Phys.*, *37* (1966) 2098.
23 J. C. Brice and G. D. King, *Nature (London)*, *200* (1966) 1346.
24 R. M. Logan and D. T. J. Hurle, *J. Phys. Chem. Solids*, *32* (1971) 1739.
25 A. J. R. de Kock, *Appl. Phys. Lett.*, *16* (1970) 100.
26 E. N. Zhelikhovskaya and L. A. Borisova, *Izv. Akad. Nauk SSSR, Neorg. Mater.*, *5* (1969) 239.
27 J. Hornstra, *J. Phys. Chem. Solids*, *5* (1958) 129.
28 P. Haasen, *Acta Metall.*, *5* (1957) 598.
29 J. D. Venables and R. M. Broudy, *J. Appl. Phys.*, *29* (1958) 1025.
30 M. S. Abrahams and C. J. Buiocchi, *J. Appl. Phys.*, *36* (1965) 2855.
31 M. S. Abrahams, J. Blanc and C. J. Buiocchi, *Appl. Phys. Lett.*, *21* (1972) 185.
32 J. P. Hirthe and J. Lothe, *Theory of Dislocations*, McGraw-Hill, New York, 1968, p. 356 *et seq.*
33 G. H. Olsen, M. S. Abrahams and T. J. Zamerowski, *J. Electrochem. Soc.*, *121* (1974) 1650.
34 R. L. Bell and A. F. W. Willoughby, *J. Mater. Sci.*, *1* (1966) 219.
35 A. Steinemann and U. Zimmerli, *Solid-State Electron.*, *6* (1963) 597.
36 E. D. Jungbluth, *Metall. Trans.*, *1* (1970) 575.
37 T. Iizuka, *J. J. Appl. Phys.*, *7* (1968) 485.
38 T. Iizuka, *J. J. Appl. Phys.*, *7* (1968) 490.
39 P. B. Hirsch, A. Howie, R. B. Nicholson, D. W. Pashley and M. J. Whelan, *Electron Microscopy of Thin Crystal*, Butterworths, London, 1965, p. 327 *et seq.*
40 G. Eckhardt, *J. Appl. Phys.*, *33* (1962) 1016.
41 H. Kressel, F. Z. Hawrylo, M. S. Abrahams and C. J. Buiocchi, *J. Appl. Phys.*, *39* (1968) 5139.
42 H. Kressel, H. Nelson, S. H. MacFarlane, M. S. Abrahams, P. Lefur and C. J. Buiocchi, *J. Appl. Phys.*, *40* (1969) 3589.
43 H. Kressel, N. E. Byer, H. Lockwood, F. Z. Hawrylo, H. Nelson, M. S. Abrahams and S. H. MacFarlane, *Metall. Trans.*, *1* (1970) 635.
44 T. Iizuka, *J. Electrochem. Soc.*, *118* (1971) 1190.
45 G. A. Rozgonyi and T. Iizuka, *J. Electrochem. Soc.*, *120* (1973) 673.
46 G. A. Rozgonyi, A. R. Von Neida, T. Iizuka and S. E. Haszko, *J. Appl. Phys.*, *43* (1972) 3141.
47 M. S. Abrahams and C. J. Buiocchi, *J. Appl. Phys.*, *37* (1966) 1973.
48 M. S. Abrahams and J. I. Pankove, *J. Appl. Phys.*, *37* (1966) 2596.
49 M. S. Abrahams, L. R. Weisberg, C. J. Buiocchi and J. Blanc, *J. Mater. Sci.*, *4* (1969) 223.
50 M. S. Abrahams and L. Ekstrom, *Acta Metall.*, *8* (1960) 654.
51 D. T. J. Hurle, *Solid-State Electron.*, *3* (1961) 37.
52 K. F. Hulme and J. B. Mullin, *Philos. Mag.*, *4* (1959) 1286.
53 C. Z. LeMay, *J. Appl. Phys.*, *34* (1963) 439.
54 W. Bardsley, J. M. Callan, H. A. Chedzey and D. T. J. Hurle, *Solid-State Electron.*, *3* (1961) 142.
55 J. J. Tietjen, M. S. Abrahams, A. B. Dreeben and H. F. Gossenberger, *Proc. 2nd Int. Symp. on Gallium Arsenide and Related Compounds, Dallas, 1968*, Inst. Phys., London, 1969, Paper 9, p. 55.
56 D. P. G. Lidbury, *D. Phil. Thesis*, University of Oxford, 1974, p. 26 *et seq.*
57 E. S. Meieran, *J. Appl. Phys.*, *36* (1965) 2544.
58 H. Steinhardt, *Ph. D. Thesis*, University of Gottengen, 1972.
59 D. Laister and G. M. Jenkins, *Solid-State Electron.*, *13* (1970) 1200.
60 M. S. Abrahams and L. Ekstrom, in H. C. Gatos (ed.), *Properties of Elemental and Compound Semiconductors*, Vol. 5, Interscience, New York, 1960, p. 225.
61 H. A. Schell, *Z. Metallkd.*, *48* (1957) 158.
62 C. S. Fuller and H. W. Allison, *J. Electrochem. Soc.*, *109* (1962) 880.
63 M. V. Sullivan and G. A. Kolb, *J. Electrochem. Soc.*, *110* (1963) 585.
64 A. Reisman and R. Rohr, *J. Electrochem. Soc.*, *111* (1964) 1425.
65 J. C. Dyment and G. A. Rozgonyi, *J. Electrochem. Soc.*, *118* (1971) 1346.
66 U.S. Patent 3,170,273 (February 23, 1965).

67 J. Knutton, personal communication, 1974.
68 R. R. Bradley, unpublished results, 1974.
69 R. D. Packard, *J. Electrochem. Soc.*, *112* (1965) 871.
70 S. Iida and K. Ito, *J. Electrochem. Soc.*, *118* (1971) 768.
71 J. G. Grabmaier and C. B. Watson, *Phys. Status Solidi*, *32* (1969) K13.
72 Yu. V. Pleskov, *Dokl. Akad. Nauk SSSR*, *143* (1962) 1399.
73 W. W. Harvey, *J. Electrochem. Soc.*, *114* (1967) 472.
74 R. W. Haisty, *J. Electrochem. Soc.*, *108* (1961) 790.
75 S. Iiyama, I. Ida and S. Furumoto, *Rev. Electr. Commun. Lab.*, *17* (1969) 1022.
76 B. A. Unvala, D. B. Holt and Aung San, *J. Electrochem. Soc.*, *119* (1972) 318.
77 R. W. Bicknell, *J. Phys. D.*, *6* (1973) 1991.
78 J. F. Nye, *Acta Metall.*, *1* (1953) 153.
79 M. S. Abrahams, *J. Appl. Phys.*, *35* (1964) 3626.
80 J. G. Grabmaier, personal communication, 1972.
81 B. C. Grabmaier and J. G. Grabmaier, *J. Cryst. Growth*, *13/14* (1972) 635.
82 J. L. Richards and A. J. Crocker, *J. Appl. Phys.*, *31* (1960) 611.
83 T. S. Plaskett and A. H. Parsons, *J. Electrochem. Soc.*, *112* (1965) 954.
84 D. E. Bolger, unpublished results, 1973.
85 E. Sirtl and A. Adler, *Z. Metallkd.*, *52* (1961) 529.
86 M. J. Cardwell and D. J. Stirland, reported at *5th Int. Symp on Gallium Arsenide and Related Compounds, Deauville, 1974.*
87 M. J. Cardwell and D. J. Stirland, to be published.
88 J. O. Hvålgard, S. L. Andersen and T. Olsen, *Phys. Status Solidi (a)*, *5* (1971) K83.
89 G. B. Larrabee, *J. Electrochem. Soc.*, *108* (1961) 1130.
90 D. C. Newton, unpublished results, 1974.
91 B. Schwartz, *J. Electrochem. Soc.*, *118* (1971) 657.
92 H. Hasegawa, K. E. Forward and H. Hartnagel, *Electron. Lett.*, *11* (1975) 53.
93 A. M. Huber and G. Champier, *Proc. 3rd Int. Symp. on Gallium Arsenide and Related Compounds, 1970*, Inst. Phys., London, 1971, p. 118.
94 D. J. Stirland and R. Ogden, *Phys. Status Solidi (a)*, *17* (1973) K1.
95 D. J. Stirland, unpublished work, 1974.
96 T. S. Noggle, B. F. Day, F. A. Sherrill and F. W. Young, Jr., *Bull. Am. Phys. Soc.*, *10* (1965) 324.
97 R. E. Reason, *Properties of Metallic Surfaces*, Inst. Metals, London, 1953, p. 327.
98 G. Nomarski and A. R. Weill, *Rev. Metall. (Paris)*, *52* (1955) 121.

*Thin Solid Films*, *31* (1976) 171–183
© Elsevier Sequoia S.A., Lausanne—Printed in Switzerland


# CRYSTALLOGRAPHIC DEFECTS IN EPITAXIAL SILICON FILMS

K. V. RAVI
*Mobil Tyco Solar Energy Corporation, 16 Hickory Drive, Waltham, Mass. 02154 (U.S.A.)*
(Received June 2, 1975; accepted June 17, 1975)

A review of the salient theories advanced to account for the formation of crystallographic defects in epitaxially deposited silicon films on silicon substrates is presented. Experimental observations of defect generation processes in epitaxial films due to slip bands and point defect complexes in the substrate are discussed with an analysis of the possible mechanisms involved. The effects of growth stacking faults on p–n junction and bipolar transistor characteristics are also discussed.

---

## 1. INTRODUCTION

Homoepitaxy of thin films of silicon on silicon is a well-established technology used extensively in the manufacture of semiconductor devices and circuits[1, 2]. The primary parameters that are typically controlled during the production of high quality epitaxial layers are the resistivity and thickness of the layer, and particular attention is paid to the uniformity of these parameters across a wafer and between different wafers. The crystallographic defect densities have been dramatically reduced to quite acceptable limits by the use of low dislocation density substrates and *in situ* HCl etching procedures prior to the deposition of the epitaxial layers. However, with the increase in area of many new devices and circuits, higher perfection epitaxial layers with lower defect densities are becoming more desirable. In addition, certain new defect generation processes have been observed to operate when low or zero dislocation density substrates and larger diameter (3 in or greater) wafers are used for the manufacture of devices and circuits involving epitaxy.

In this paper a brief review of the various theories advanced to explain the generation of crystallographic defects in silicon epitaxial layers is presented. This is followed by experimental observations of some new defect generation processes with a discussion of the possible mechanisms involved. A concluding section briefly reviews reported electrical effects of defects in epitaxial films, with particular emphasis on epitaxial or growth stacking faults.

## 2. EPITAXIAL LAYER QUALITY

The principal crystallographic defects that have been observed in silicon epitaxial films on silicon substrates are dislocations and growth stacking faults. The synthesis technique used was $SiO_4$ decomposition.

*2.1. Dislocations*

The incorporation of dislocations into epitaxial layers can occur by essentially two methods. The most commonly observed process is the propagation of dislocations in the substrate into the growing epitaxial layer. In the growth of epitaxial films of compound semiconductors using liquid phase epitaxy, it has been observed that the dislocation density is often significantly lower in the epitaxial layers than in the substrate[3–5]. A number of models have been proposed to account for this phenomenon. One proposal[6] suggests that straight dislocations acquire jogs as they propagate into the epitaxial layer, causing changes in the growth direction which could eventually result in the dislocations growing parallel to the film–substrate interface or in the formation of dislocation loops close to the interface. The formation of the jogs is postulated to be a result of the action of point defects in the vicinity of the dislocation(s). Other effects that have been suggested to account for the propagation of dislocations into the epitaxial layer are the "catalytic influence" of metallic solvents[3] and stress reduction and orientation of the dislocations[4]. The morphology of the dislocations in the substrate has been demonstrated to influence their propagation into the epitaxial film. In the case of Ga(AsP) epitaxy it is found that inclined dislocations tend to propagate whereas a reduction in the density of straight misfit dislocations is observed[5,7]. Although this phenomenon has principally been observed in compound semiconductors, recent studies[8] of epitaxial deposits on ribbon silicon substrates grown by the EFG process[9] have shown that there is a significant reduction in the dislocation densities in thick epitaxial films compared with the relatively high defect densities found in the ribbon substrates.

A model recently proposed[10] to account for the reduction in dislocation density in epitaxial films assumes that epitaxial growth occurs by the lateral propagation of ridges parallel to the substrate surface. The lateral growth is postulated to drag dislocations along, hence changing their direction, which results in propagation parallel to the interface or in loop formation due to annihilation reactions between dislocations.

Although dislocation propagation from the substrate into the epitaxial layers is possible, this mode of defect incorporation is not very significant in view of the increasing tendency to use dislocation-free substrates. An alternative method by which dislocations can be generated in epitaxial layers results from the introduction of mechanical stresses into the crystals during processing operations associated with epitaxy for device applications. Further discussion of this effect will be presented later.

*2.2. Growth stacking faults*

By far the greatest amount of work on defects in epitaxial layers has been devoted to the study of growth stacking faults[11]. These defects have been shown to form as closed figures or as ribbons bound by Shockley partial dislocations with fault vectors of $\frac{1}{6}a <112>$. Most observers report the nucleation of these defects at the substrate–epitaxy interface, so a relationship between the size of defect and the epitaxial layer thickness can be established. Indeed a rough estimate of epitaxial layer thickness can be obtained from a measure of the width of epitaxial faults. A number of theories have been advanced to account for the origin of epitaxial stacking faults in silicon. The salient features of each of the theories will be discussed briefly.

Principally three factors influence the generation of growth stacking faults* in silicon epitaxial layers. These can be broadly classified into (a) the effects of contaminants introduced into the epitaxial reactor during film deposition, (b) the effects of contaminants and mechanical damage on the substrate surface and (c) the effects of crystallographic defects existing in the substrate close to the surface of the crystal.

*2.2.1. Effects of gaseous contaminants*

Hydrocarbon contamination during film growth has been demonstrated to nucleate stacking faults[12–14]. The presence of carbon in gaseous form has been shown to generate various defects in silicon films, including twins, stacking faults and the formation of polycrystalline layers when the carbon concentrations reach a relatively high level (of the order of $10^{19}$ atoms $cm^{-3}$). Studies of thin film growth by Abbink *et al.*[15] and Cullis and Booker[16] have shown that growth occurs by the propagation of two-dimensional steps on the surface in directions normal to the thickness direction. The carbon atoms are postulated to be swept along by the lateral growth layers, slowing down growth. New nucleation occurs on top of the uncompleted layers resulting in the occupation of carbon atoms in substitutional sites in the silicon lattice and the concomitant nucleation and propagation of stacking faults. A similar model by Dyer[17] suggests that twin formation in vapor grown silicon crystals is a result of the incorporation of carbon or SiC in the form of short chains.

*2.2.2. Effects of surface contaminants and damage*

Contaminants on the substrate surface have been observed to nucleate stacking faults in the deposited epitaxial layers[13]. Oxide patches on the surface have been implicated specifically, although other contaminants such as organic films can function in a similar way. Finch *et al.*[11] suggest that patches of oxide can produce steps at the substrate surface, the height of which may not be equal to an integral multiple of the lattice spacing in the growth direction. Deposition of silicon layers is considered to occur with maintenance of coherence at the growth interface and across oxide patches, the coherence being retained by the formation of stacking faults.

Mechanical damage on the substrate surface in the form of scratches, saw marks etc. also results in stacking fault nucleation[11]. Slip bands generated in the substrate by prior heat treatments, such as those associated with sub-collectors or buried $n^+$ layer diffusions, can also nucleate stacking faults in the epitaxial films. This may be due to the surface damage causing incorrect deposition of the first atomic layers of the films as a result of the geometric displacement produced by surface steps on the first deposited layers. Mechanical damage in the substrate can result in the generation of a high density of shallow dislocation loops[18] which

* Generally, other defects such as tri-pyramids, raised triangles etc. are also observed in association with growth stacking faults, and the nucleation mechanisms of all these defects are considered to be similar.

can nucleate stacking faults because of the dissociation of the dislocations into partials enclosing faults which propagate into the growing epitaxial layer.

*2.2.3. The effects of defects in the substrate*

By far the most interesting phenomenon is associated with the influence of bulk crystallographic defects in the substrates on the formation of growth stacking faults.

The presence of second phases such as $SiO_2$ and SiC in the bulk silicon close to the surface can initiate faults in the epitaxial films. Booker and Stickler[18] propose that fault nucleation sites are extrinsic partial dislocation loops in the substrate, and Unvala and Booker[19] have detected dislocation loops at the substrate epitaxial interface using transmission electron microscopy. Jaccodine[20] has proposed that vacancy clusters in the substrate collapse into stacking fault tetrahedra which then initiate faults in the epitaxial films. These clusters are presumed to be a result of growth mistakes, gross imperfections, oxide islands or any other form of lattice mismatch. Vacancy clusters in the substrate formed as a consequence of the particular conditions existing during crystal growth can influence the formation of growth stacking faults[21].

Many of the models just discussed have been demonstrated to be operative in silicon epitaxy. In the next section two processes are discussed, with experimental observations, that are particularly relevant to the increasing use of low dislocation density large-area substrates.

## 3. FAULT NUCLEATION ON DISLOCATION-FREE SUBSTRATES

### *3.1. Mechanical stress induced slip*

High rates of insertion and withdrawal of wafers into and from the hot zone of diffusion and oxidation furnaces are known to cause plastic deformation. This is because thermal gradients are established between the periphery and centers of the crystals, which induce sufficient stress to cause dislocation nucleation and propagation[22].

Oxidation and diffusion operations prior to epitaxy are frequent occurrences in the fabrication of many integrated circuits. Consequently, wafers containing a high density of slip induced dislocations and slip bands can result prior to epitaxy. This effect, however, can be avoided if suitable precautions are taken to ensure uniform heating and cooling of the wafers.

Another way in which plastic deformation can occur in circular wafers is as a result of non-uniform heating during epitaxy. Central and peripheral slip has been shown to occur in slightly curved silicon wafers during epitaxy[23]. This effect can also result from the presence of hot spots on the susceptor in the epitaxial reactor. As would be expected, the presence of any stress raisers in the form of sharp notches etc. around the wafer periphery can magnify the problem of slip.

Plastic deformation during film deposition can result in stacking fault nucleation due to the creation of surface steps as a result of slip. If the stress due to wafer curvature, hot spots on the susceptor or the presence of stress raisers at the wafer edges is sufficiently high, a continuous generation of dislocations can occur with the result that both film deposition and dislocation generation and motion proceed concurrently. This could lead to a continuous or steady state nucleation of growth stacking faults, which would imply that fault nucleation can also occur on freshly deposited silicon layers and not only at the substrate–epitaxial layer interface. Examples of this effect are shown in Figs. 1 and 2. Figure 1(a) shows slip bands generated in a <111> oriented wafer; the slip became evident after epitaxy (and a preferential etch). Figures 1(b) and 1(c) are high magnification displays showing stacking faults, mostly in the form of closed triangles, and dislocations in the form of dark dots (due to insufficient resolution at these magnifications). Faults are observed to form along the slip bands and vary in size, indicating non-common nucleation points. Figure 2 shows a similar

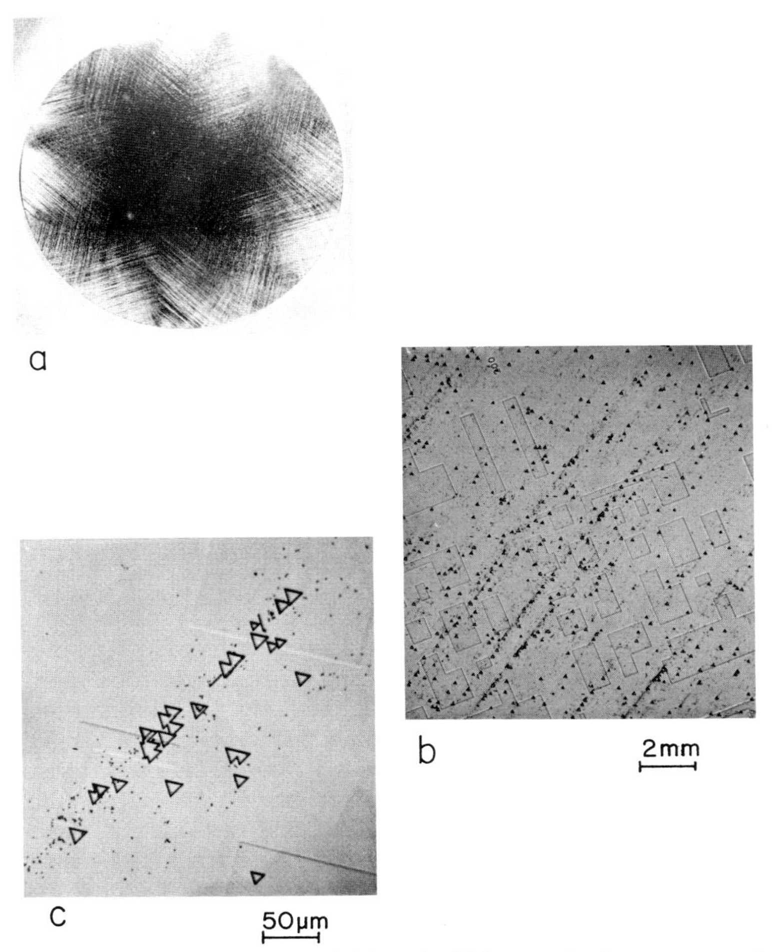


Fig. 1. (a) Slip bands generated in the epitaxial layer and substrate as a result of thermal stress. (b) and (c) High magnification micrographs demonstrating the nucleation of growth faults along slip bands.

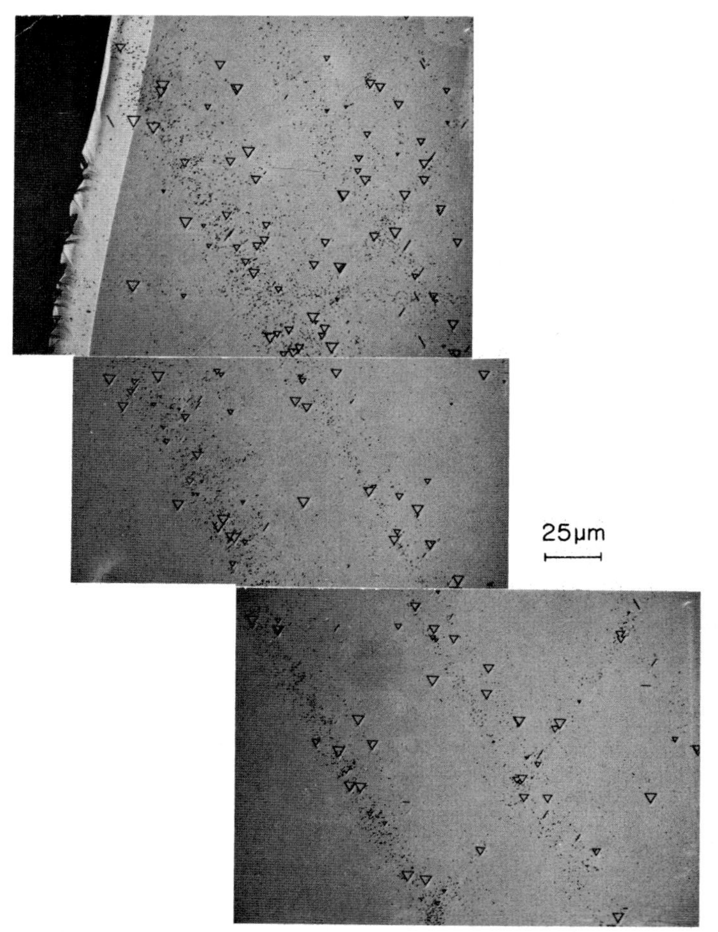


Fig. 2. Stacking fault nucleation along slip bands, the slip being caused by the presence of mechanical damage at the wafer edges. The variation in size of faults indicates non-common nucleation points.

case where slip has been initiated at regions of mechanical damage at the wafer periphery, the driving force for activating and moving the dislocations being the temperature associated with the epitaxial process. The faults are aligned along slip bands (<110> direction) and are again observed to vary in size, which lends credence to the hypothesis that fault nucleation is a steady state process in which new faults nucleate during the deposition process while the faults nucleated early in the deposition process or at the substrate–epitaxial layer interface continue to propagate. The problem of slip during epitaxy is of greater consequence for larger diameter wafers owing to the greater difficulty of maintaining wafer flatness. Also the influence of the quality of the wafer edges on dislocation nucleation is an important effect. Carefully prepared wafer edges that are rounded by mechanical or chemical processes minimize the possibility of stress raisers and hence of dislocation generation.

*3.2. Nucleation at point defect complexes*

The presence of point defect complexes in dislocation-free silicon has been shown to cause nucleation of stacking faults during oxidation[24], and to influence the precipitation kinetics of metals such as copper and lithium[25] and of oxides such as the oxides of silicon[26]. These complexes have been postulated to be due to the association of vacancies and oxygen atoms in the silicon crystals, the association occurring during crystal growth[25]. Some evidence exists to demonstrate that the complexes can nucleate growth faults in epitaxial layers[21].

The distribution pattern assumed by the vacancy clusters is often in the form of a spiral or branded pattern across the wafer surface[24] and is associated with re-melt effects due to crystal rotation[25]. These clusters or complexes can be detected by preferential etch techniques and by X-ray techniques following copper decoration[25]. Preferential etching generally results in the formation of shallow etch pits distributed in this banded or spiral pattern, generally called a "swirl" pattern[27].

Factors such as rotation rates, growth rates, crystal diameter and crystal growth ambient strongly influence the formation and distribution of impurity complexes. When wafers are cut from the crystals, a finite density of these complexes intersect the surface and can influence the quality of deposited epitaxial layers. This effect is shown in Figs. 3 and 4. Figures 3(a) and 3(b) are dark and bright field optical micrographs of a portion of a wafer following epitaxy (with no preferential etch). Defects distributed in discrete bands are evident. By preferentially etching the opposite side of the substrate it was established that the distribution of banded defects in the epitaxial film is similar to the distribution of point defect complexes in the substrate. The higher magnification micrograph in Fig. 4 shows that the defects generated in the epitaxial layer are growth stacking faults. The defects are confined to bands with almost defect-free zones between the bands. The influence on the nucleation of growth stacking faults of vacancy complexes in the substrate is exactly analogous to their influence on the nucleation of oxidation-induced stacking faults[24].

The specific manner in which impurity complexes in the substrate nucleate growth stacking faults in the epitaxial film is a matter of conjecture at this time. However, three possible mechanisms suggest themselves as the most likely processes by which this form of defect generation can proceed. One mechanism is the direct effect of vacancy clusters on the perfection of the initially deposited layers of the thin films. Since a complex of point defects, such as vacancies near the surface, represents a local disturbance in the lattice, the deposited epitaxial films can be perturbed locally and generate growth faults. Also, as mentioned above, vacancy complexes nucleate extrinsic stacking faults when subjected to oxidation. These faults are observed to nucleate at the wafer surface as well as at regions beneath the surface in the bulk of the crystal[24]. An example of this is shown in Fig. 5. The banded distribution of stacking faults, formed as a result of oxidation, is evident. The distribution pattern assumed by these defects is identical to that of point defect complexes in the unoxidized crystal. Oxidation steps such as those associated with sub-collector diffusion in integrated circuit processing can nucleate stacking faults at vacancy complexes in the crystal. These faults

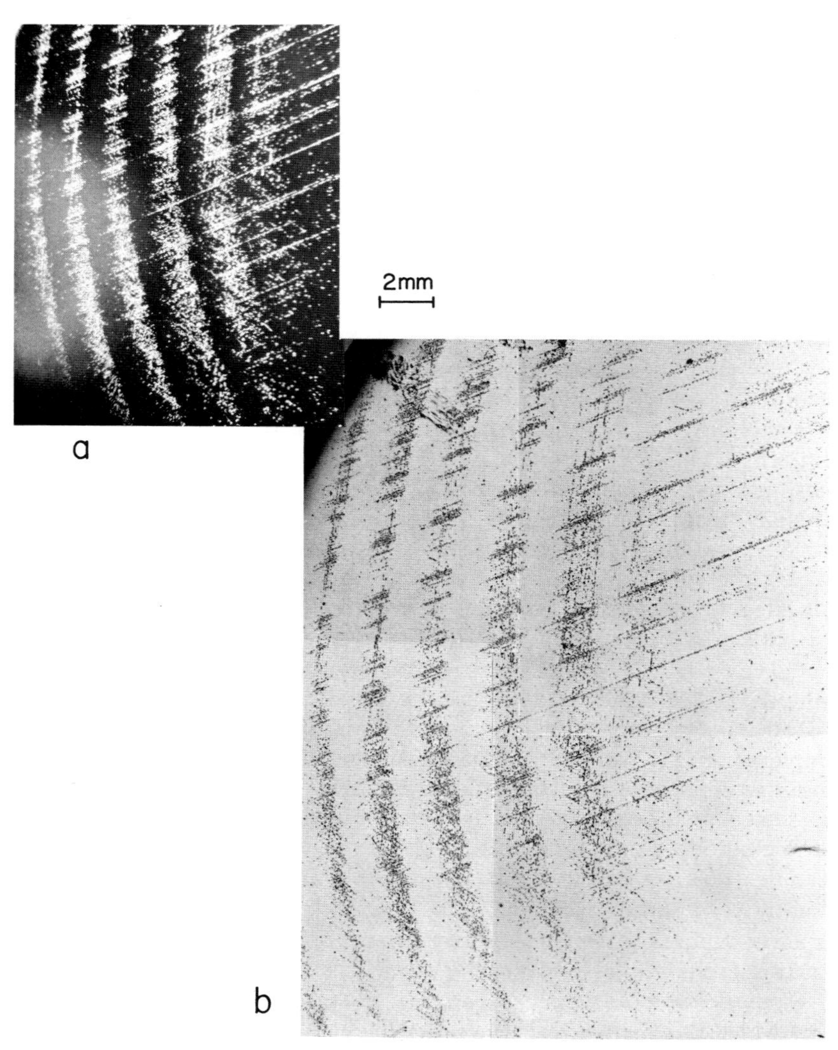


Fig. 3. Dark and bright field optical micrographs of the banded distribution of growth stacking faults.

in turn can nucleate growth stacking faults in the subsequently deposited epitaxial layers. The presence of point defect complexes in the crystal can also promote the nucleation of second phases in the form of metallic precipitates and oxides. Extended oxidation of silicon has been demonstrated to result in the formation of discrete oxide precipitate clusters in dislocation-free crystals containing a distribution of point defect complexes[26]. The localization of metallic and oxide precipitates at discrete sites across the wafer surface, and throughout the bulk, can influence the quality of epitaxial films by functioning as nucleation sites for growth stacking faults.

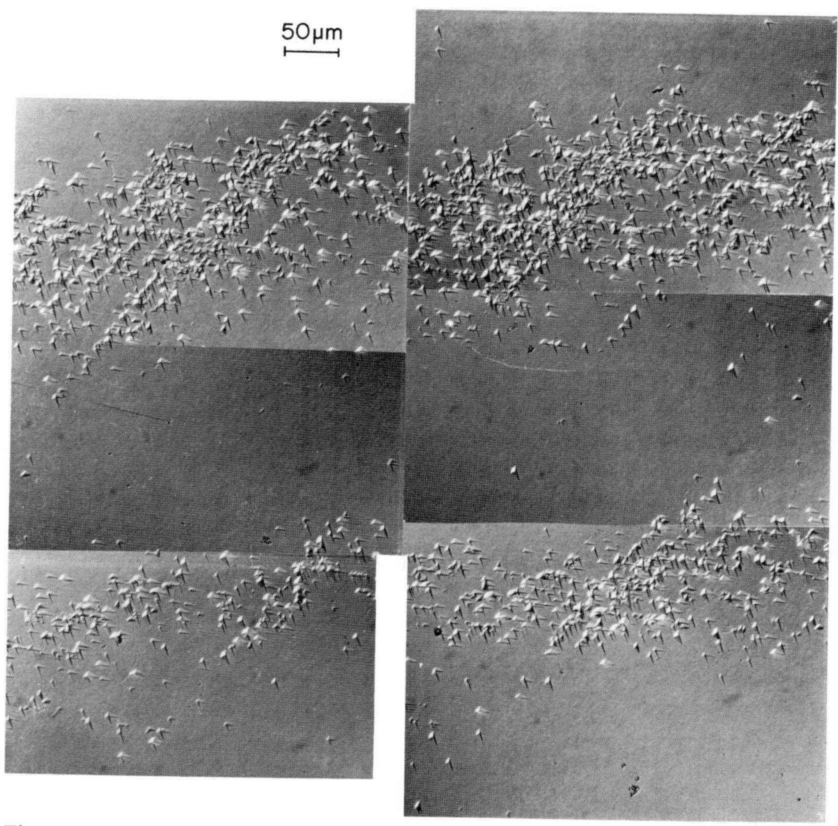


Fig. 4. The heterogeneous nucleation of growth stacking faults along "swirl" bands. Note that the regions between bands are almost free of defects.

The influence of bulk defects such as point defect complexes on epitaxial layer quality is such that *in situ* HCl etching procedures generally do not affect the defect nucleation process; this is due to the distributed nature of these grown-in defects. Removal of layers of silicon by HCl gas should result in the exposure of fresh defects at the surface so that growth stacking faults can be nucleated irrespective of the particular surface preparation steps employed prior to film deposition.

It should be noted that because the epitaxial layer and the substrate had comparable doping levels (1–5 Ω cm), no misfit dislocations were observed in the samples discussed here.

## 4. ELECTRICAL EFFECTS

Two effects specifically have been associated with epitaxial stacking faults. These are the effects of stacking faults on the reverse leakage and breakdown

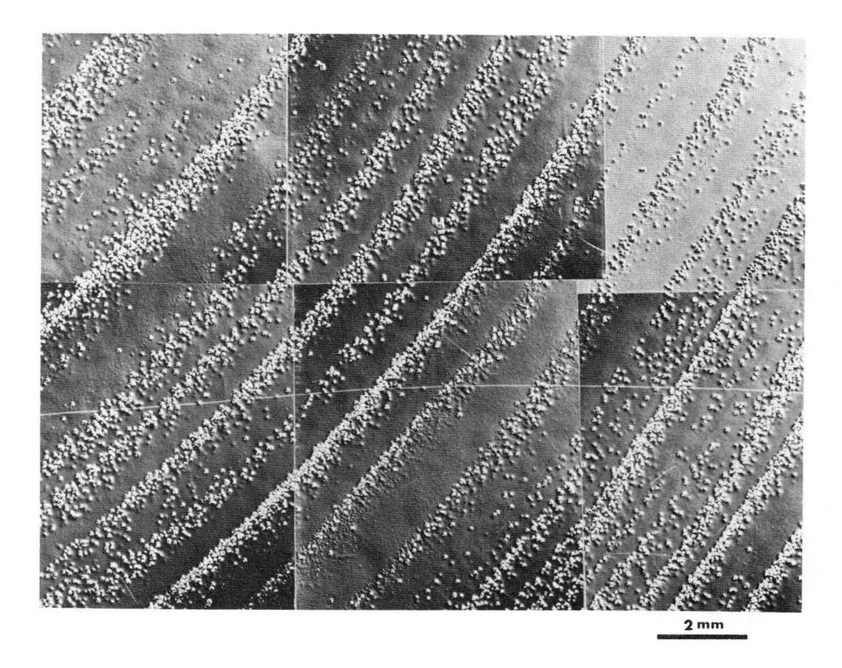


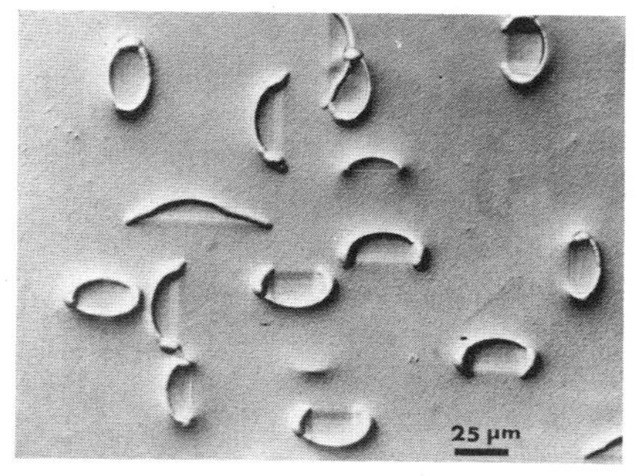


Fig. 5. Banded distribution of oxidation induced stacking faults in a wafer containing a non-random distribution of point defect complexes ("swirl"). Note the similarity in the distribution of growth stacking faults in Figs. 3 and 4 in epitaxial films grown on similar substrates.

characteristics of p–n junctions and the formation of emitter–collector pipes or shorts in bipolar transistors.

The introduction of excess reverse currents in p–n junctions has been associated with the stair-rod dislocation at the corners of growth faults. The early work of Queisser and Goetzberger[28] demonstrated the formation of microplasmas at epitaxial faults in p–n junctions, the microplasmas occurring when the junctions were reverse biased. In addition, a "softening of the knee" in the reverse *I–V* characteristic was also observed.

The effects of growth faults in bipolar transistors are of greater technological significance in view of their influence on the performance characteristics and yield of such devices. Barson *et al.*[29] have demonstrated that electrical pipes or shorts occurring in double diffused transistors fabricated in epitaxial films can be correlated with epitaxial faults. Based on an analysis employing optical techniques, scanning and transmission electron microscopy, a similar conclusion was reached by Varker and Ravi[30] in wide base-width transistor structures with epitaxially grown p- and n-type base layers.

One of the reject electrical modes observed in such structures exhibits the *I–V* characteristics of typical electrically piped double diffused transistors. These characteristics are manifested by excess emitter–collector leakage currents, in a variety of three terminal measurements, and a voltage coupling effect which is evident on open-circuit voltage measurements of the emitter or collector terminals. Using the electron beam induced current mode of operation (EBIC)[31, 32] of the SEM, a correlation was obtained between the electrical characteristics of the transistor and local regions of current enhancement in the emitter. Optical techniques were used to detect the presence of microplasma sites at the surface of partially fabricated structures by applying a reverse bias to the substrate epitaxial junction. Surface etching and copper decoration techniques revealed that the current enhancement sites as observed in the SEM and the optical microplasma sites were associated with stacking faults. These were subsequently identified as growth stacking faults by using transmission electron microscopy. To verify that the faults originate at the substrate and terminate at the epitaxial film surface the transistors were cross-sectioned through the stacking faults and metallographically examined after suitable bevel and strain procedures. The cross sections indicated that the faults did originate at the epitaxial layer–substrate interface and extended through the film ending at the surface. The surface etching technique produced strong differential etching effects at the faulted region, suggesting impurity decoration of the faults and an apparent contouring of the epi–substrate p–n junction. Furthermore, small amounts of phosphorus were detected in the faulted region of the n–p–n structure with an X-ray microprobe analyzer.

Based on these observations a model was proposed wherein epitaxial faults which nucleate at the substrate surface act as impurity sinks during growth and hence promote out-diffusion of phosphorus or boron from the substrate to produce autodoping of the faulted region. The preferential accumulation of phosphorus or boron in the fault can occur as a result of the combined effects of the formation of a Cottrell atmosphere around the partial dislocations bounding the fault and a possible Suzuki interaction of the impurity atoms with the faulted region. An n- or p-type channel can develop during the growth process and extend towards the epitaxial surface, providing a conductive path which is evident in the electrical measurements after emitter diffusion. The sketches in Fig. 6 illustrate the mechanism.

Recently Kato *et al.*[33] have observed epitaxial faults in the SEM using the EBIC technique. Their observations show that a reduction in the induced current signal occurs in the region near the stair-rod dislocations. By using an X-ray apparatus in the SEM they detected the presence of phosphorus in the faults, which supports the hypothesis discussed above. In addition, the broad region of contrast around the partial dislocations indicates that recombination effects

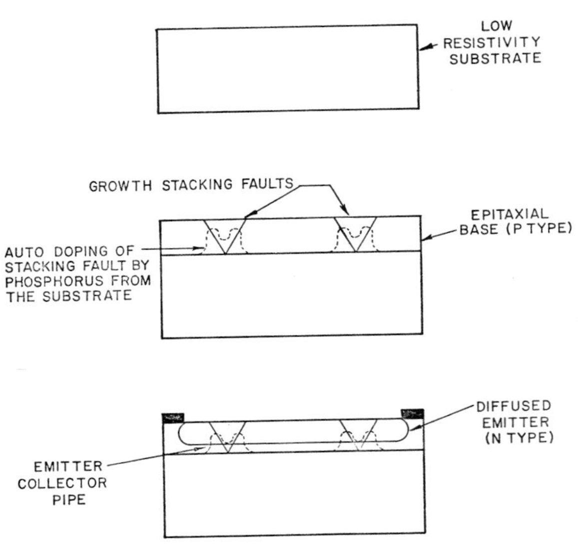


Fig. 6. Sketches illustrating a mechanism whereby growth stacking faults can cause electrical pipes in wide base-width bipolar transistors.

associated with the partials extend to distances considerably in excess of the dimensions of the dislocations. Similar effects have been observed in oxidation-induced stacking faults in silicon wherein an extended zone of local high density recombination–generation centers are observed around the defects[34]. On examining the effects of dislocations on the properties of MOS capacitors, McCaughan and Wonsiewicz[35] found a similar effect wherein each dislocation appears to have approximately the same effect on the $C$–$V$ curves as $10^4$ charges $cm^{-2}$ at the interface. Since $10^4$ atoms at the interface are not directly affected by the dislocation, an electrically active zone extending over large areas around the dislocation and caused by local impurity segregation can explain the effect.

The electrical effects of growth stacking faults appear to be very similar to those of other crystallographic defects such as dislocations and oxidation-induced stacking faults, the electrical activity being related largely to defect–impurity association. The distinction between growth faults and other defects may lie in the property of growth faults of extending through the epitaxial or active layer in which the junctions are formed. Dislocations and, in particular, oxidation-induced stacking faults are not necessarily subjected to this restriction since a large density of such defects are confined to near-surface regions[36]. In such situations the electrical effects due to the defects are a function of the depth of the p–n junction, whereas epitaxial faults can influence shallow and deep junctions equally.

REFERENCES

1 P. Rai-Choudhury, in H. R. Huff and R. R. Burgess (eds.), *Semiconductor Silicon*, Electrochem. Soc., Princeton, N. J., 1973, p. 243.

2 J. Bloem, in H. R. Huff and R. R. Burgess (eds.), *Semiconductor Silicon*, Electrochem. Soc., Princeton, N.J. 1973, p. 180.
3 G. A. Wolff and B. N. Das, *J. Electrochem. Soc.*, *113* (1966) 299.
4 R. S. Mroczkowski, A. F. Witt and H. C. Gratos, *J. Electrochem. Soc.*, *115* (1968) 545.
5 M. S. Abrahams, L. R. Weisberg, C. J. Buiocchi and J. Blanc, *J. Mater. Sci.*, *4* (1969) 233.
6 R. H. Saul, *J. Electrochem. Soc.*, *118* (1971) 793.
7 M. S. Abrahams, L. R. Weisberg and J. J. Tiejen, *J. Appl. Phys.*, *40* (1969) 3758.
8 H. Kressel, P. Robinson, S. H. McFarlane, R. V. D'Aiello and V. L. Dalal, *Appl. Phys. Lett.*, *25* (1974) 197.
9 H. E. Bates, F. H. Cocks and A. I. Mlavsky, *Proc. 9th IEEE Photovoltaic Specialists Conf.*, *1972*, IEEE, New York, 1972.
10 H. J. Queisser, *J. Cryst. Growth*, *17* (1972) 169.
11 R. H. Finch, H. J. Queisser, G. Thomas and J. Washburn, *J. Appl. Phys.*, *34* (1963) 406.
12 G. R. Booker and B. A. Joyce, *Philos. Mag.*, *14* (1966) 301.
13 P. Rai-Choudhury, *J. Electrochem. Soc.*, *118* (1971) 7.
14 Y. Avigal and M. S. Wieber, *J. Cryst. Growth*, *9* (1971) 127.
15 A. C. Abbink, R. M. Broudy and C. P. McCarthy, *J. Appl. Phys.*, *39* (1968) 4673.
16 A. G. Cullis and C. R. Booker, *J. Cryst. Growth*, *9* (1971) 132.
17 L. D. Dyer, *J. Electrochem. Soc.*, *118* (1971) 987.
18 G. R. Booker and R. Stickler, *Appl. Phys. Lett.*, *3* (1963) 158.
19 B. A. Unvala and G. R. Booker, *Philos. Mag.*, *1* (1966) 11.
20 R. J. Jaccodine, *Appl. Phys. Lett.*, *2* (1963) 201.
21 D. I. Pomerantz, *J. Electrochem. Soc.*, *119* (1972) 255.
22 H. R. Huff, R. C. Bracken and S. N. Rea, *J. Electrochem. Soc.*, *118* (1971) 143.
23 L. D. Dyer, H. R. Huff and W. W. Boyd, *J. Appl. Phys.*, *42* (1971) 5680.
24 K. V. Ravi and C. J. Varker, *J. Appl. Phys.*, *45* (1974) 263.
25 A. J. R. deKork, *Appl. Phys. Lett.*, *16* (1970) 100.
26 K. V. Ravi, *J. Electrochem. Soc.*, *121* (1974) 1090.
27 K. V. Ravi and C. J. Varker, *Appl. Phys. Lett.*, *25* (1974) 69.
28 H. J. Queisser and A. Goetzberger, *Philos. Mag.*, *8* (1963) 1063.
29 F. Barson, M. S. Hess and M. M. Roy, *J. Electrochem. Soc.*, *116* (1969) 304.
30 C. J. Varker and K. V. Ravi, *Spring Meeting Electrochem. Soc.*, *Houston*, *1972*, Extended Abstract No. 62.
31 W. Czaja and J. R. Patel, *J. Appl. Phys.*, *36* (1965) 7476.
32 K. V. Ravi, C. J. Varker and C. E. Volk, *J. Electrochem. Soc.*, *120* (1973) 533.
33 T. Kato, H. Koyama, T. Matsukawa and R. Shimizu, *J. Appl. Phys.*, *45* (1976) 3732.
34 C. J. Varker and K. V. Ravi, *J. Appl. Phys.*, *45* (1974) 272.
35 D. V. McCaughan and B. C. Wonsiewicz, *J. Appl. Phys.*, *45* (1974) 4982.
36 C. J. Varker and K. V. Ravi, in H. R. Huff and R. R. Burgess (eds.), *Semiconductor Silicon*, Electrochem. Soc., Princeton, N.J., 1973, p. 724.

*Thin Solid Films*, *31* (1976) 185–216



# X-RAY CHARACTERIZATION OF STRESSES AND DEFECTS IN THIN FILMS AND SUBSTRATES

G. A. ROZGONYI AND D. C. MILLER
*Bell Laboratories, Murray Hill, N. J. 07974 (U.S.A.)*
(Received May 26, 1975; accepted June 17, 1975)

The development of semiconductor devices requiring an increasingly complex sequence of epitaxial, dielectric and metallic layers has generated the need for precise measurements of the stresses and defects formed during the processing of these multilayer structures. The imaging of native and process-induced defects in semiconductors using X-ray topography has become a well-established procedure in the electronics industry. Recent advances in the quantitative determination of the layer stress simultaneously obtained during the X-ray imaging process yield a combined approach which opens up new possibilities for defect analysis in electronic materials. The present report reviews the X-ray topographical approach, both for transmission and reflection, with special emphasis on defect and strain field imaging. Quantitative stress analysis and automatic Bragg angle control techniques are then described. Representative examples are given for each individual topic to illustrate how the separate techniques work. Finally, a discussion is presented of how all the techniques can be used in a complementary fashion to solve a broad range of both fundamental and technologically relevant problems.

## 1. INTRODUCTION

Recent advances in microelectronics and technology have resulted in the development of complex multilayer structures consisting of alternating layers of metals, dielectrics and semiconductors grown on single-crystal substrates. Also, the physical size of the layers has been reduced, not only in lateral area but also in the thickness direction, *e.g.* integrated optical systems require as many as seven alternating single-crystal layers each about 1 μm thick, and each with a different composition, while superlattice structures may have a hundred or more sub-micron layers. New technologies such as ion implantation have created buried layer structures which must be characterized in relation to the host crystal. The interactions between the layers and the cumulative effect on the substrate during the course of high temperature processing steps or long-term device operation near room temperature cannot always be predicted in advance and often result in failure of the device. Thus there is a clear need for a variety of precise electrical, chemical and structural information about these thin film materials systems.

Materials scientists have responded to this need with an alphabet soup of sophisticated analytical tools such as AES, ISS, DIMA, STEM etc., several of which are the subject of other articles in this special issue, and some of which cost a large fraction of a million dollars to install. However, in the area of non-destructive defect analysis of thick specimens the basic tools have not changed significantly since Lang described his scanning transmission X-ray diffraction camera in 1958[1, 2]. What has changed is the varied way in which the Lang camera has been applied to problems of stress and associated elastic and plastic deformation phenomena. In the present article we describe several of these developments with particular emphasis on applications to thin film electronics technology. The sensitivity of X-ray diffraction to lattice distortions and the ability of X-rays to penetrate thick samples present a unique situation for the simultaneous non-destructive analysis of such intimately related data as the layer stress and the substrate defect density.

As an introduction to the variety of information that is revealed in a transmission X-ray topograph we present in Fig. 1(a) an enlarged area of a sample of interest to engineers processing MOS integrated circuits. The sample was fabricated by oxidizing a silicon wafer, removing the oxide from half the wafer and then depositing stripes of W over the boundary between Si alone and Si covered with $SiO_2$, as shown in Fig. 1(b). Basically, what we have is a scaled-up structural model of an MOS device. It is found that as-deposited W films are continuous and are elastically accommodated by the substrate. However, after heat treatment for 1 h at 850 °C, if the Si is not protected with $SiO_2$ the W reacts with it. Note that the W stripes have peeled off to the right of the Si–$SiO_2$ boundary in Fig. 1. Also, the local stresses produced by the W–Si interactions are relieved, in part, by the generation of large dislocation loops (arrows D) leaving slip traces (arrow S) which penetrate several millimetres into the $SiO_2$-protected regions. The stresses around pinholes and regions of poor contact in the W yield a black on white contrast feature (arrow P). Note that the vertical line between the $SiO_2$ and Si halves of the wafer is white, which, as will be explained below, is indicative of compressively stressed layers. Since the white line is considerably wider at the edge of the W stripes, we know that the stress is higher in the W stripes than in the $SiO_2$ layer.

Therefore, in this single X-ray image it is possible to analyze the effect of annealing on the introduction of bulk defects, on layer–layer interactions, on layer uniformity and coherence, and on the sign and relative magnitude of layer stresses before and after the heat treatment. Adding quantitative values of the layer stresses, which are also obtained using the X-ray camera, to the qualitative information in the topograph enables a device processing engineer to control, or at least to be aware of, the level of stress and plastic flow due to a specific procedure.

On first being presented with an X-ray image of the sort shown in Fig. 1(a) four questions are invariably asked.

(1) How was the image recorded, *i.e.* what is an X-ray topographic camera?

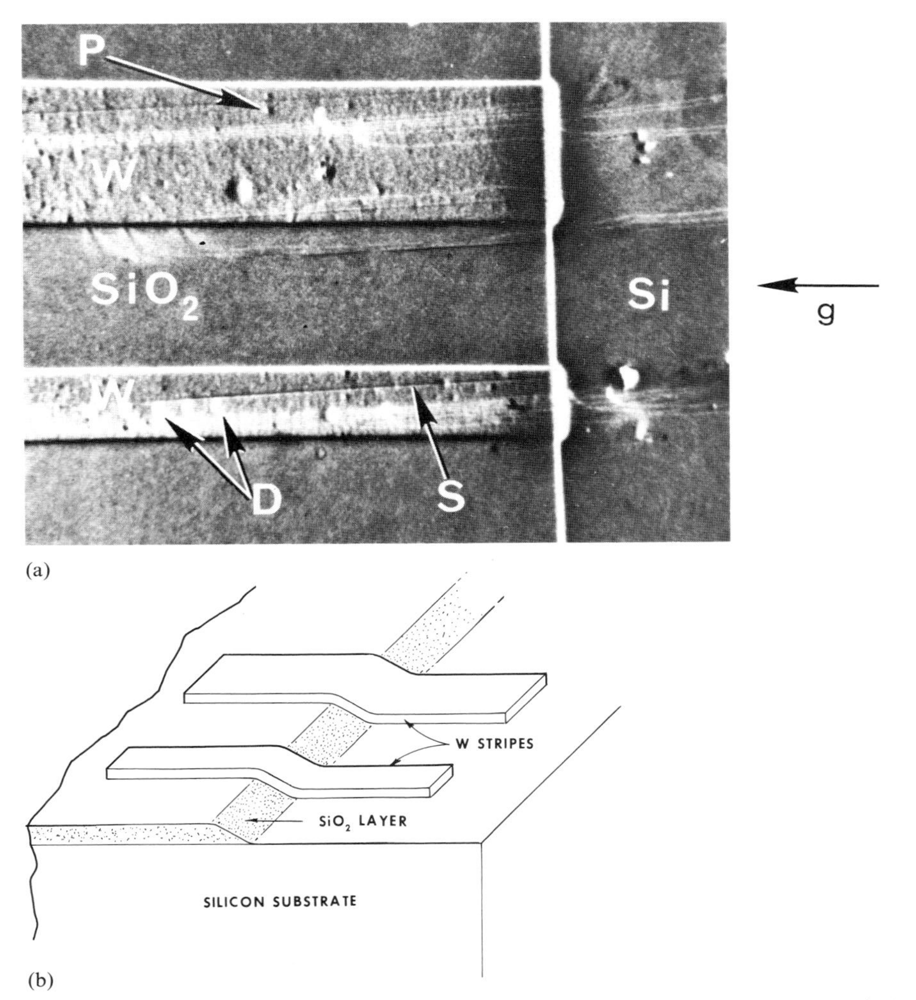


Fig. 1. (a) Transmission X-ray topograph of a silicon wafer with tungsten stripes deposited over the boundary between the $SiO_2$-covered and bare (100)-oriented silicon substrate. The arrows P, D and S point to pinholes, dislocation loops and accompanying slip traces, respectively. The W stripes are misaligned about 10° with respect to the diffraction vector $\boldsymbol{g}$. Dynamical contrast mode. (b) Schematic diagram of W stripes and $SiO_2$ coverage on a silicon sample.

(2) Why does the X-ray intensity, *i.e.* contrast, vary at defects and strain centers to give black or white lines on a grey background?

(3) What additional information, besides imaging, does the topograph yield about the imperfections in a sample?

(4) How can this technique be applied to various materials science problems?

In the present report we hope to answer these questions by describing the qualitative aspects of the basic diffraction phenomena; how dislocations and strain gradients are imaged; the equipment and how it can be modified to provide quantitative stress data on thin film/substrate combinations; plus an extensive series of representative applications intended not only to illustrate how these techniques work, but also to emphasize the broad range of both fundamental and technologically relevant problems that can be solved. The scope of this article will be limited to those measurements which can be performed on a commercially available Lang camera. Although the authors have added certain electronic control features to their own equipment, these functions can be performed manually to obtain essentially the same data. Stress and defect studies on a Lang camera require that the substrate, but not necessarily the layer or layers deposited on the substrate, be a single crystal.

## 2. TRANSMISSION X-RAY TOPOGRAPHY

### 2.1. *Basic principles*

Several extensive review articles[3-6] dealing with the experimental details of X-ray topography, the theoretical basis of the diffraction phenomena involved and the application of topography to semiconductor technology appeared in 1970. The salient features of these articles, which are almost exclusively related to transmission X-ray topography, will be outlined here in order to establish a framework within which to discuss applications made in the last five years; in addition the combined reflection, transmission and quantitative stress approach favored by the present authors will be discussed.

In its simplest form only three pieces of equipment are necessary to set up a facility for measuring stress and imaging defects. They are: an X-ray generator to supply the photons which will probe the sample; a Lang camera to collimate the X-rays, orient the sample and mechanically scan the sample under study; and finally, a means to image and identify defects and lattice distortions.

A schematic diagram of how this is accomplished on a Lang camera is shown in Fig. 2. The horizontal divergence of the incident beam is limited by the slits to about 4′ of arc. In this way the two images created by the characteristic $K\alpha_1$–$K\alpha_2$ doublet can be separated by the crystal. The vertical divergence is easily controlled by using a point source of X-rays and adjusting the distance between source and sample. Thus the sample is irradiated by a very narrow, but tall, beam which passes through the sample from A to B. The crystal is then set, according to Bragg's law, so that lattice planes (*hkl*) normal to the entrance and exit faces will diffract as shown schematically in Fig. 2. Other diffracting planes can be chosen but the principles remain the same.

The diffracted beam exits from the back of the sample and passes through a slit in a screen which prevents the transmitted beam and fluorescent radiation from blackening the photographic emulsion. The emulsion is exposed to a beam A′B′ yielding an image of a section AB of crystal. If a defect or strain center D is located in the crystal between A and B, the X-rays interact with the local lattice distortions to create intensity variations, which are observed as contrast at D′ on the photographic plate. This yields a "section" topographic image which can be used to locate in depth the position of imperfections within the crystal. An example of an array of interfacial misfit dislocations in an epitaxial silicon sample analyzed

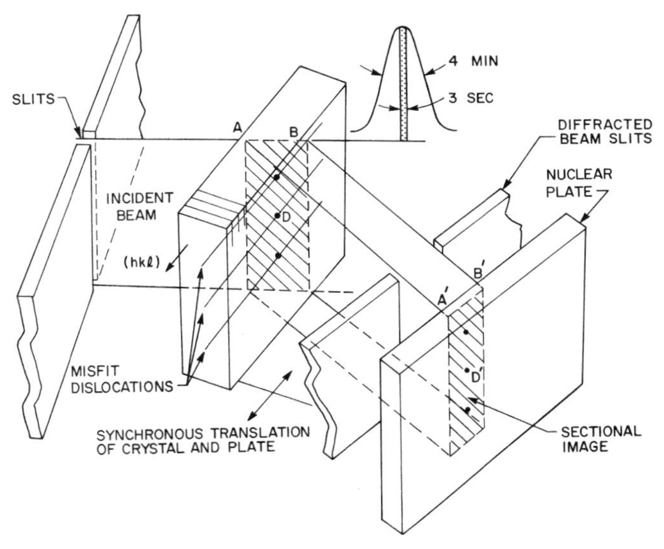


Fig. 2. Schematic representation of the defect imaging process by Lang X-ray topography.

by McFarlane as part of a study of high voltage devices[7] is shown in Fig. 3(b). The dislocations are separated by about 120 μm at a depth of 240 μm and are viewed end-on in the section topograph. In order to image the entire length of the misfit dislocations it is necessary to scan the sample and the photographic plate synchronously so that the entire sample intercepts the narrow beam. The image thus obtained is called a "projection" or scanning transmission X-ray topograph and gives the imperfection content of the entire sample, as shown in Fig. 3(a). Note that the analysis of defects in the thickness direction (obtained with the section topograph) has been sacrificed in order to survey the overall imperfection density of the entire volume of the wafer.

*2.2. Contrast formation*

The topographs in Figs. 3(a) and 3(b) both show dislocations as black images on a grey background. This type of contrast is easily explained by comparing the amount of the incident beam energy which is "dynamically" diffracted by a perfect crystal with that "kinematically" diffracted by an imperfect or mosaic crystal. It was mentioned above that the horizontal divergence of the incident beam is typically about 4′ of arc; however, a perfect crystal lattice will only select several seconds of arc of this beam for diffraction, as shown schematically in Fig. 2. A defect-free crystal will produce a uniformly grey featureless image on the photographic plate. In an imperfect crystal, the bent atomic planes surrounding a defect will locally diffract more of the incident beam energy from the "wings" of the divergent incident beam, thus producing a black image of the defect on a grey background, as shown by the dislocations in Figs. 3(a) and 3(b) which are diffracting kinematically.

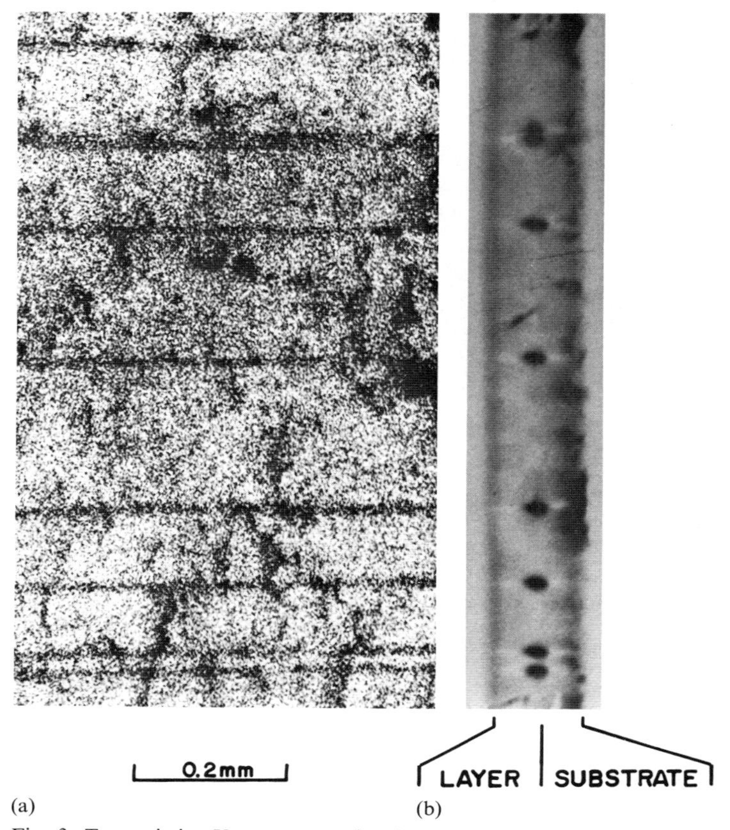


(a) (b)

Fig. 3. Transmission X-ray topographs of a sample with misfit dislocations and kinematical contrast mode as illustrated in Fig. 2: (a) scanning projection topograph; (b) section topography. (By courtesy of S. H. McFarlane.)

This simple black on grey contrast analysis should be compared with the more complicated black and white on grey features discussed in the topograph presented in Fig. 1(a). Explanation of the black on white contrast requires more explicit understanding of the dynamical interactions of the wave fields within the crystal, which are briefly discussed next. The kinematical and dynamical theories of diffraction are distinguished in the following way. The kinematical theory assumes that the incident and diffracted beams pass in and out of the crystal without mutually interacting, suffering only normal absorption according to exp $(-\mu t)$, where $\mu$ is the linear absorption coefficient and $t$ is the thickness penetrated. This assumption is valid in practice for highly imperfect or mosaic crystals; for defects in thin (*i.e.* $\mu t \leqq 1$) nearly perfect crystals in the transmission mode, for which Fig. 3 is an example; and for defects near the surface of crystals of any thickness in the reflection mode, in which the X-rays enter and exit from the same surface.

The dynamical theory takes into account mutual interactions of the incident and diffracted beams and must be used to obtain a complete description of the diffracted beam intensities, especially in perfect crystals for which $\mu t > 1$. For thick crystals with $\mu t > \sim 2$ the image contrast due to the kinematically diffracted beam is reduced due to absorption. However, in low-defect material Borrmann[8] discovered that portions of the dynamical beam will be transmitted through very thick crystals. The expression "anomalous" transmission, or "Borrmann effect", is used when referring to this mode since $\mu t$ may be greater than 20. Penning and Polder[9] have described the trajectories of this low absorption mode as a flow of energy along the $(hkl)$ lattice planes, as shown by the arrow C in Fig. 4(a). On reaching the exit face of the crystal the anomalously transmitted beam separates into diffracted (D) and forward diffracted (FD) components whose intensities are equal in an undistorted region of the sample (see wave C). An image of either of the two diffracted beams will produce a uniformly grey topograph in a perfect crystal. However, the lattice curvature at the edge of a compressively stressed layer deposited onto the perfect substrate will change the relative intensities of the two beams locally. This creates an enhanced diffracted beam intensity at B, where the curvature is in the direction of the diffracted beam, and a reduction at A, where the lattice curvature favors the forward diffracted beam. Imaging the diffracted beam yields black and white lines at B and A, respectively, relative to the grey background produced by waves like C. If a layer stressed in tension is considered, as shown in Fig. 4(b), the overall curvature is of opposite sense, as are the intensity modulation effects on the diffracted beams.

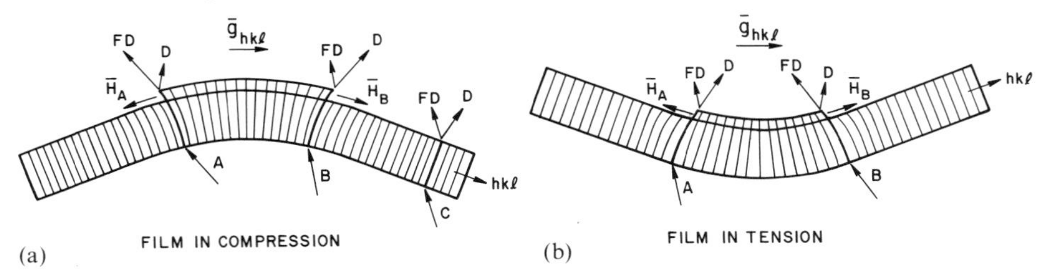


Fig. 4. Schematic representation of the origins of black and white on grey contrast for crystals with large $\mu t$ bent by layers stressed in (a) compression and (b) tension.

The black and white on grey strain contrast is illustrated by the tungsten stripes in Fig. 1(a), which are white at the top edge and black at the bottom. This black and white contrast depends on the local substrate lattice curvature and the direction of the diffraction vector $\boldsymbol{g}$ and can be used to determine the sense of the layer stress. Note that the white line is much wider at the right-hand edge of a W stripe and that this edge is more nearly perpendicular to the diffraction vector $\boldsymbol{g}$, which is tilted about 10° to the horizontal and is parallel to the slip traces (see arrow S in Fig. 1(a)). If the W stripes were perfectly aligned along $<110>$ then the black on white contrast would disappear from the top and bottom of the stripes since the (220) set of lattice planes participating in the diffraction process would not be locally distorted. Using a $(\bar{2}20)$ diffraction vector would, of course, reverse the role of the vertical and horizontal edges. These contrast relationships and their application to stresses in deposited thin films are summarized by the $\boldsymbol{g}\cdot\boldsymbol{H}$ criteria in Table I. The unit vector $\boldsymbol{H}$ is defined as having a direction which is always pointing away from that edge of the layer which is being studied. For diffused, implanted or locally damaged structures $\boldsymbol{H}$ would likewise point away from the volume of the substrate so treated. In Fig. 1(a) the right and left portions of the wafer are separated by a vertical white line and $\boldsymbol{g}\cdot\boldsymbol{H}<0$. These conditions tell us that the $SiO_2$ and W layers are both compressively stressed, which in the case of the $SiO_2$ is due to its having a larger thermal expansion coefficient than the substrate, leading to a convex warpage when viewed from the film side.

TABLE I

CONTRAST CRITERIA FOR DETERMINING LAYER STRESS

| | *Topograph contrast** | *Stress* |
|---|---|---|
| $\boldsymbol{g}\cdot\boldsymbol{H}>0$ | black | compressive |
| $\boldsymbol{g}\cdot\boldsymbol{H}<0$ | white | compressive |
| $\boldsymbol{g}\cdot\boldsymbol{H}>0$ | white | tensile |
| $\boldsymbol{g}\cdot\boldsymbol{H}<0$ | black | tensile |

* In certain cases the contrast may be strong or weak, *i.e.* black and dark grey on a light-grey background, depending on the value of $\mu t$ and the magnitude of the local deformation.

### 2.3. *Defect contrast*

The lattice distortions associated with dislocations also have a directional character which is described by the Burgers vector $\boldsymbol{b}$. The distortions interact with the X-ray beam in much the same way as in the case of the strain generated by oxide stripes discussed above. Therefore, by suitable arrangements of $\boldsymbol{g}$ and consideration of other diffraction conditions, such as the line direction $\boldsymbol{u}$ of the dislocation, it is possible to determine the direction and magnitude of the Burgers vector which yields the particular type of dislocation, whether screw, edge or mixed. A thorough analysis of all the contrast details of dislocations is quite involved and beyond the scope of this paper, but discussion of the topographs shown in Figs. 5(a) and 5(b) will serve to illustrate several of the main points.

The X-ray topographs (dynamical contrast mode) of Fig. 5 contain a wealth of information about the origins and character of interfacial misfit dislocations introduced during the LPE growth of a GaAlAsP layer on a (001) GaAs substrate and will be discussed in greater detail in Section 5. At this point we wish to discuss only the effects of local displacements and lattice curvature on the visibility criteria for the dislocations A and B, which are 60° dislocations with a mixed edge and screw character, and the dislocation C, a pure edge dislocation. Note that by using orthogonal diffracting planes to obtain two images of the same area of the crystal the interfacial misfit dislocations may exhibit: (1) strong contrast, like A and B in Fig. 5(a) and C in Fig. 5(b); (2) weak contrast, as A and B in Fig. 5(b); or (3) no contrast at all. C is invisible in Fig. 5(a) and can only be located because each end of the defect changes its line direction in order to exit from the top surface of the sample (see the two dislocation segments $C_f$ which are visible in Fig. 5(a)).

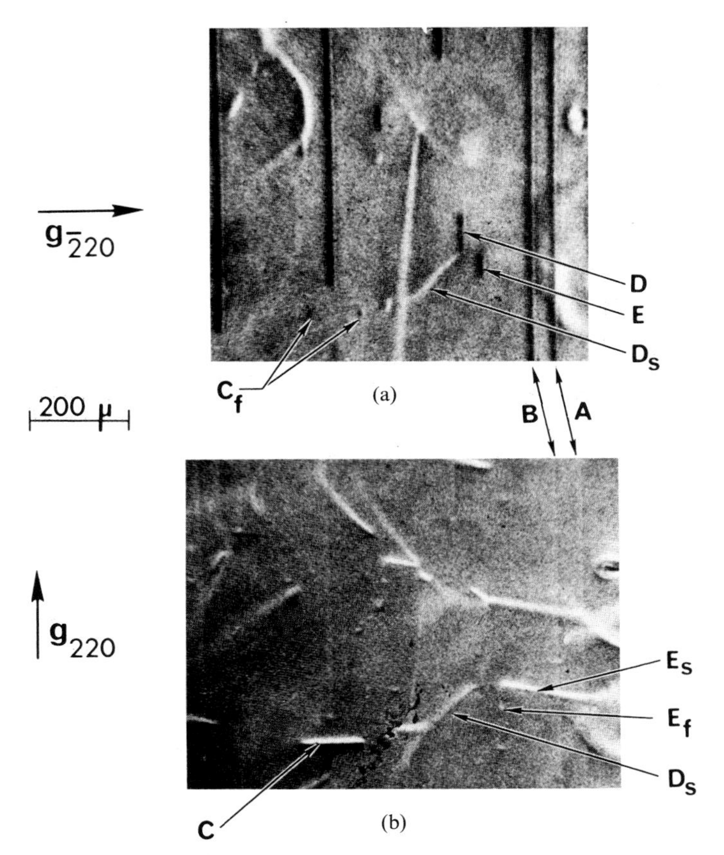


Fig. 5. Transmission X-ray topographs of a heteroepitaxial GaAlAsP/GaAs sample with interfacial misfit dislocations, arrows A–E. D and E originate at threading substrate dislocations $D_S$ and $E_S$, C is connected to the top film surface via segments $C_f$, while A and B are entirely contained by the substrate–layer interface.

The explanation of these contrast effects lies in an understanding of the anisotropic bending of lattice planes surrounding a line defect. For example, the crystal planes containing the Burgers vector of a screw dislocation have essentially no distortion, while the planes normal to the Burgers vector have maximum distortion. Therefore a topograph of a crystal containing a screw dislocation made by diffraction off those planes containing the Burgers vector, *i.e.* $\boldsymbol{g} \cdot \boldsymbol{b} = 0$, will show no contrast at the image of the dislocation. However, a topograph made with the highly distorted planes normal to the Burgers vector, *i.e.* $|\boldsymbol{g} \cdot \boldsymbol{b}| = 1$, will show maximum contrast. Similar rules apply for edge dislocations except that for complete extinction or loss of visibility to occur it is also necessary that $\boldsymbol{g} \cdot (\boldsymbol{b} \times \boldsymbol{u}) = 0$, where $\boldsymbol{u}$ is a unit vector along the line of the dislocation. The various conditions applicable to A, B and C in Fig. 5 are summarized in Table II. Note that 60° mixed dislocations can never be extinguished because of the complementary behavior of the edge and screw components.

TABLE II

X-RAY TOPOGRAPH IMAGE CONTRAST FOR MISFIT DISLOCATIONS IN THE (001) LAYERS OF FIG. 5 FOR $<220>$ DIFFRACTION VECTORS $\boldsymbol{g}$, $\frac{1}{2}a <110>$ BURGERS VECTORS $\boldsymbol{b}$, AND DISLOCATION LINE DIRECTIONS $\boldsymbol{u}$

| *Topograph image* | *Dislocation character* | | |
|---|---|---|---|
| | *90° edge* | *0° screw* | *60° mixed* |
| Visible:<br>$\|\boldsymbol{g}\cdot\boldsymbol{b}\| > 0$ | C in Fig. 5(b) | Screw components A and B in Fig. 5(b) | Always visible, edge component in Fig. 5(a) and screw in Fig. 5(b) |
| Not visible:<br>$\boldsymbol{g}\cdot\boldsymbol{b} = 0$ plus<br>$\boldsymbol{g}\cdot(\boldsymbol{b}\times\boldsymbol{u}) = 0$ for edge | C in Fig. 5(a) edge component of A and B in Fig. 5(b) | No pure screws in Fig. 5. Component vanishes for A and B in Fig. 5(a) | Never vanishes |

A final comment on Fig. 5 is that the 60° dislocations show marked variations in the nature of the black on white contrast for the same value of $\boldsymbol{g}\cdot\boldsymbol{b}$ depending on the relative orientation of $\boldsymbol{g}$ and the direction of the dislocation $\boldsymbol{u}$. (Compare the segments D and $D_s$ of the same dislocation in Fig. 5(a).) Further discussion of misfit dislocation contrast effects can be found in recent papers by Mader *et al.*[14], who did a comparison electron microscopy analysis, and Kishino and Ogirima[15], who observed screw-type misfit dislocations.

## 3. REFLECTION TOPOGRAPHY

### *3.1. Basic principles*

In addition to the transmission or Laue geometry, where the X-rays enter one surface of the crystal and exit from another surface, X-ray topographs can also be obtained by diffracting the X-rays back out of the same surface they have entered. This is called the reflection, Berg–Barrett[16–17] or Bragg geometry and is shown schematically in Fig. 6. The symbols will be discussed in some detail below. Reviews devoted to the operational aspects of reflection topography can be found in refs. 18 and 19. At this point we wish to emphasize that in the reflection geometry only the outer skin of a sample is imaged, typically 2–10 μm and occasionally down to a depth of 50 μm. Reflection topographs are therefore

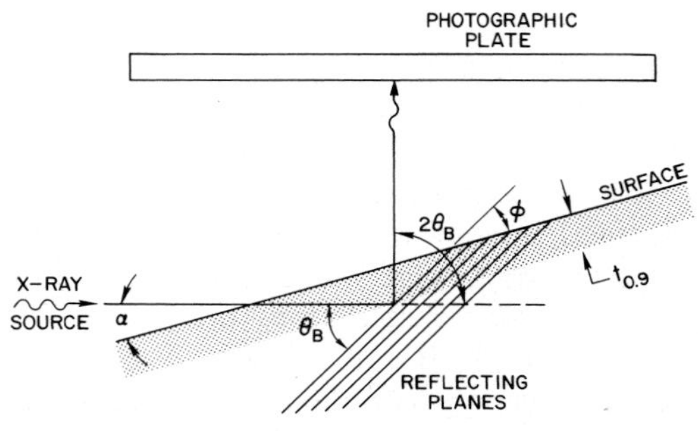


Fig. 6. Schematic diagram of sample alignment for reflection X-ray topography.

particularly well suited to studies of defects in epitaxial layers and to materials-processing problems where surface strain is an important consideration.

Reflection topography of semiconductor samples has not been as popular as the transmission mode because the transmission images have better contrast since both dynamical and kinematical images may be present, whereas in the reflection mode the contrast is entirely kinematical. Also, the entire length of the bulk defects inclined to a surface are imaged in the transmission case while only those segments exiting near the surface can be revealed in a reflecting topograph, thus reducing the volume of the strained material which contributes to contrast formation. This is illustrated in Figs. 7(a) and 7(b) where several large dislocation loops in the same sample have been imaged by both techniques. Note that the depth of the loops is such that the reflection topograph only sees portions of the segments threading up to the surface (arrows $L_2$). The common origin of the loop $L_2$ could not be deduced from the reflection XRT alone.

However, there are many situations where selectively imaging the surface defect structure is advantageous; *e.g.*: where the dislocation density is such that counting dislocation emergence sites is simpler than analyzing overlapping lines; where extended dynamical black on white images at the edges of the device

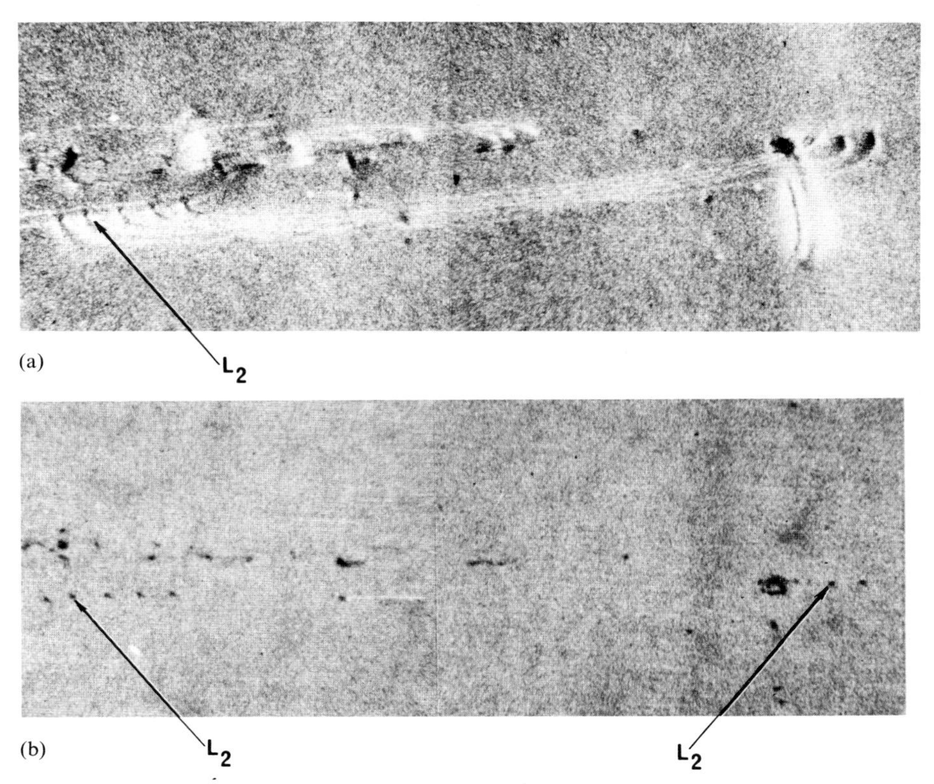


Fig. 7. Comparison of (a) transmission and (b) reflection X-ray topographs of the same dislocation loops.

patterns are not desirable; and where an independent analysis of heteroepitaxial layers is required. Also, the substrate is often so highly damaged that the individual dislocations cannot be resolved because of too much absorption in the anomalous transmission case or the diffraction of essentially all of the divergent incident beam in the thin crystal case, *i.e.* no intensity or no contrast, respectively. We will therefore attempt to present the various reflection XRT approaches developed in the past few years and show how they often complement the transmission topographs.

The choice of the diffracting plane is an important consideration in the analysis of crystal surfaces by reflection topography. The ability to image dislocations and/or surface strain gradients will depend on the depth of penetration of the incident X-rays, the magnitude of the distortions in the vicinity of the defects, and the angular range of reflection for regions of perfect crystal, called the Darwin–Prins halfwidth[20]. Other factors such as geometrical image distortions and the diffracted beam intensity[21] must also be evaluated. All these variables are coupled through the (*hkl*) plane which produces the image. A very useful analysis and compilation of data for all possible (*hkl*) for six semiconductor materials (Si, Ge, GaP, GaAs, InAs and InSb) has been presented by Halliwell *et al.*[22], for wafers with surface orientations of {100}, {110} or {111}. Portions of their results are reproduced in modified form as Table III.

The relationship between the geometric factors is illustrated in Fig. 6. Experimentally, crystals can be readily set up on a conventional Lang camera or, if a Lang camera is not available, on a modified goniometer head[18, 19, 21] fitted with a film cassette.

Operationally, the angle of incidence $\alpha$ should be small for maximum illumination of the sample and minimum interference with the photographic plate; the penetration depth $t_{0.9}$ should be larger than the epi-layer thickness, and the diffracted beam angle $2\theta_B$ should be near 90° for minimum distortion of the image. However, as pointed out by Halliwell *et al.*[22], the X-ray image contrast will depend on the difference between the intensities diffracted from perfect and defective crystal regions. This depends on the size and severity of the strain and the effective volume of the material around a dislocation, or strain center, for which the Bragg angle has deviated by more than $\Delta\theta$. Therefore, reflecting planes with small $\Delta\theta$ lead to better strain-field sensitivity and topographs with improved contrast.

An illustration of the trade-offs that are available for 10 μm thick homoepitaxial layers of Si, GaP or GaAs on (100)-oriented substrates can be made with Table III. If we consider a (422) reflection, $\alpha$ and $2\theta_B$ are ideal for all three materials, while for silicon $t_{0.9}$ is twice the layer thickness and $\Delta\theta$ is 7.5″. For GaP the (422) reflection would just barely sample the interface of a layer 10 μm thick, whereas GaAs interfacial defects would not be imaged at all due to the shallow penetration. Also, $\Delta\theta$ would have increased to 20″, further degrading the image quality. With GaAs and GaP it is therefore wiser to choose a (511) reflection and work with a somewhat larger $\alpha$ but a much improved $\Delta\theta$ and adequate sample penetration. Other considerations, such as the desirability of imaging etch pits, scratches etc., would be favored by choosing the smallest $\alpha$, whereas Burgers vector studies of

TABLE III

DIFFRACTION DATA FOR REFLECTION TOPOGRAPHY OF (100) SURFACES USING Cu K$\alpha$ RADIATION

| *Reflecting planes (hkl)* | *Inclination to (100) surface* | *Silicon, $\mu = 141$ cm$^{-1}$* | | | | *Gallium phosphide, $\mu = 294$ cm$^{-1}$* | | | | *Gallium arsenide, $\mu = 494$ cm$^{-1}$* | | | |
|---|---|---|---|---|---|---|---|---|---|---|---|---|---|
| | $\phi$ | $\theta$ (deg) | $\alpha$ (deg) | $\Delta\theta$ (s) | $t$ (μm) | $\theta$ (deg) | $\alpha$ (deg) | $\Delta\theta$ (s) | $t$ (μm) | $\theta$ (deg) | $\alpha$ (deg) | $\Delta\theta$ (s) | $t$ (μm) |
| 311 | 25.3 | 28.1 | 2.8 | 12.4 | 7.5 | 28.0 | 2.7 | 24.6 | 3.5 | 26.9 | 1.6 | 39.4 | 1.5 |
| 400 | 0 | 34.6 | 34.6 | 3.5 | 46.3 | 34.5 | 34.5 | 6.3 | 22.2 | 33.0 | 33.0 | 8.3 | 15.5 |
| 422 | 35.3 | 44.0 | 8.7 | 7.5 | 21.5 | 43.9 | 8.6 | 13.4 | 10.2 | 41.9 | 6.6 | 20.0 | 5.9 |
| 511 | 15.8 | 47.5 | 31.7 | 2.6 | 53.9 | 47.3 | 31.5 | 5.0 | 25.8 | 45.1 | 29.3 | 6.2 | 17.9 |
| 440 | 45 | 53.4 | 8.4 | 7.3 | 20.7 | 53.1 | 8.1 | 12.9 | 9.7 | 50.4 | 5.4 | 20.4 | 4.9 |
| 531 | 32.3 | 57.1 | 24.8 | 3.1 | 48.1 | 56.8 | 24.5 | 5.8 | 23.0 | 53.7 | 21.4 | 7.3 | 15.2 |
| 620 | 18.5 | 63.8 | 45.3 | 3.6 | 67.5 | 63.5 | 45.0 | 6.3 | 32.3 | 59.5 | 41.0 | 7.9 | 22.4 |
| 533 | 40.3 | 68.5 | 28.2 | 3.4 | 51.4 | 68.0 | 27.7 | 6.3 | 24.5 | 63.3 | 20.0 | 7.6 | 15.9 |
| 335 | 62.8 | 68.5 | 5.7 | 6.6 | 14.3 | 68.0 | 5.2 | 12.8 | 6.4 | 63.3 | 0.5 | 45.7 | 0.5 |

$\theta_B$ is the Bragg angle; $\phi$ the inclination of various (*hkl*) reflecting planes to the surface; $\alpha = \theta_B - \phi$ is the angle between the incident radiation and the surface; $\mu$ is the linear mass absorption coefficient; $\Delta\theta$ the angular range of reflection for a perfect crystal (Darwin–Prins halfwidth); and $t_{0.9}$ is the penetration depth for 90% absorption due to photoelectric absorption. (Adapted from ref. 22.)

individual dislocations would dictate the choice of several diffraction conditions to satisfy the visibility criteria, possibly at the expense of the overall image quality.

From the contrast point of view Halliwell *et al.*[22] conclude that strong dislocation contrast is obtained for Si and GaP for the majority of available reflections, while for Ge and GaAs choices are limited to those where strain sensitivity is high; for InAs and InP the high absorption and shallow penetration depths dictate the use of more energetic radiations, such as molybdenum. However, the surface morphology contrast increases with the absorption coefficient and is best for a small $\alpha$. The comprehensive tables given in ref. 22 are an indispensable aid for anyone working with reflection topography, as will be shown in the applications to be discussed.

### 3.2. *Compositional topography*

#### 3.2.1. *Depth variations*

Another variation on the reflection topography of epitaxial layers can be implemented if the lattice parameters of the layer and the substrate are sufficiently different, as is the case with heteroepitaxial layers. The lattice mismatch will give rise to different diffraction angles for the different layers as shown in the rocking curve of Fig. 8, which was first discussed by Howard and Dobrott[23], who coined the term "compositional topography". Howard and Dobrott[23] pointed out that independent images could be obtained for the layers and for the substrate as shown in the conceptual diagram reproduced in Fig. 9. Epi-layers A and B, *e.g.* $GaAs_{1-x_1}P_{x_1}$ and $GaAs_{1-x_2}P_{x_2}$ with $x_1 < x_2$, are grown on a substrate of GaAs, *i.e.* $x = 0$. In Fig. 9(a) the incident beam $I_o$ strikes the crystal at an angle $\theta$ such that only the GaAs substrate is properly oriented. The beam penetrates to a depth C below the substrate surface and a volume $V_1$ is imaged on the topographic plate. As the sample and plate are translated the volume element $V_1$ is swept across the entire substrate, thereby mapping the substrate defects. If the sample is rotated to a new Bragg condition $\theta_1$, the volume element $V_2$ in the A layer diffracts as shown in Fig. 9(b). Subsequent translation enables the A layer to be imaged independently of both the B layer and the substrate. The perfection of the B layer can be examined in a similar fashion (Fig. 9(c)).

Reflection topographic images of each of the peaks of Fig. 8 are shown in Figs. 10(a), 10(b) and 10(c) and reveal a hillock formation problem in the (Ga,As,P) heteroepi-structure. The topograph of the GaAs substrate shows the hillocks as triangular regions of zero contrast (Fig. 10(a)) due to local misorientation and enhanced absorption in the vicinity of the hillocks. The XRT of the $GaAs_{0.67}P_{0.33}$ layer (Fig. 10(b)) also shows no contrast at the hillocks and indicates that the layer morphology is very irregular and defective. Finally, the sample was oriented to diffract from a $GaAs_{0.85}P_{0.15}$ layer, which was known to be present from the rocking curve of Fig. 8. The resulting XRT (Fig. 10(c)) only imaged the hillock defects. Therefore the hillocks are probably generated at the substrate interface and contain 15% GaP. The final layer of composition 33% GaP forms around the hillocks and frequently grows over them, explaining the poor quality of the top layer and the various sizes of the hillock images in Fig. 10(c).

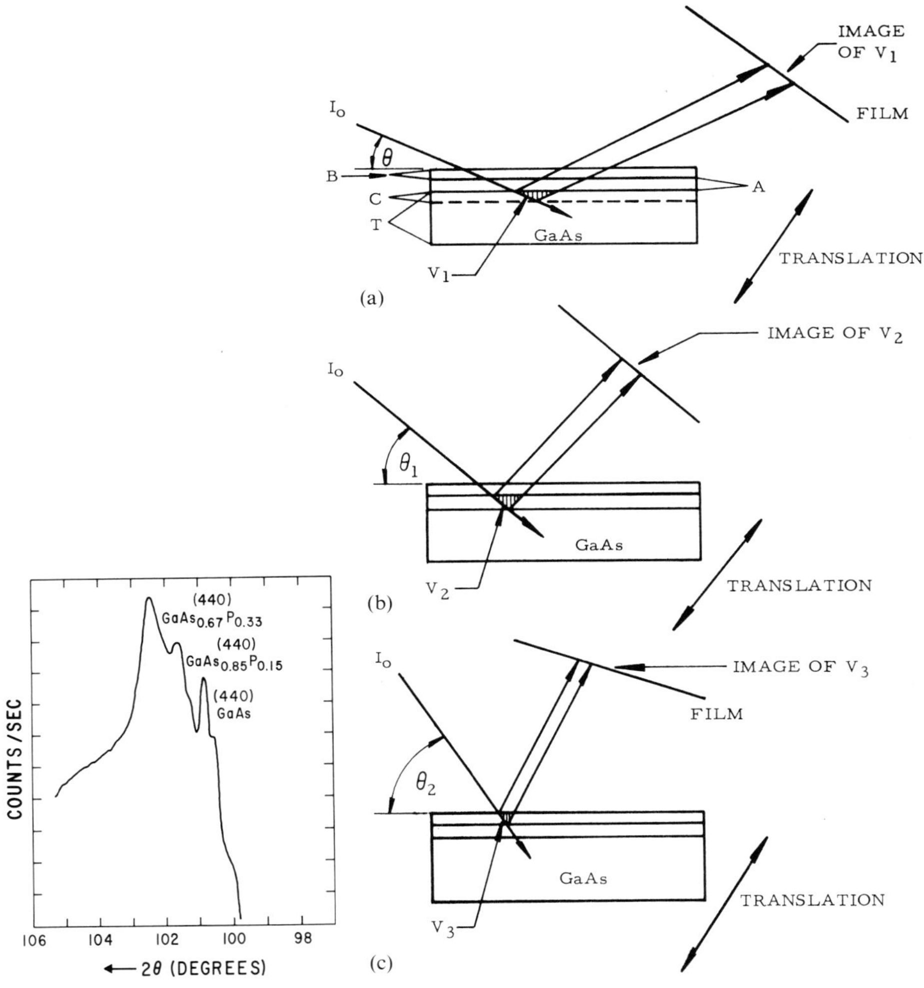


Fig. 8. Rocking curve, intensity *vs.* twice the Bragg angle, for a sample containing two GaAsP epitaxial layers on a GaAs substrate. (By courtesy of J. K. Howard.)

Fig. 9. Schematic diagram of compositional topographic analysis of the sample whose rocking curve is shown in Fig. 8. (By courtesy of J. K. Howard.)

Certain limiting conditions discussed by Howard and Dobrott for the use of compositional topography are:

(1) the total film thickness must be less than the X-ray penetration depth,

(2) the diffraction angles of the film and substrate must be sufficiently separated to ensure exact selection of $\theta$, and

(3) the absorption coefficient must be constant for each layer, *i.e.* $x$ must be reasonably uniform.

The penetration depth and image quality can be optimized using tables[22] and are not a problem for most group III–V materials less than 10 μm thick.

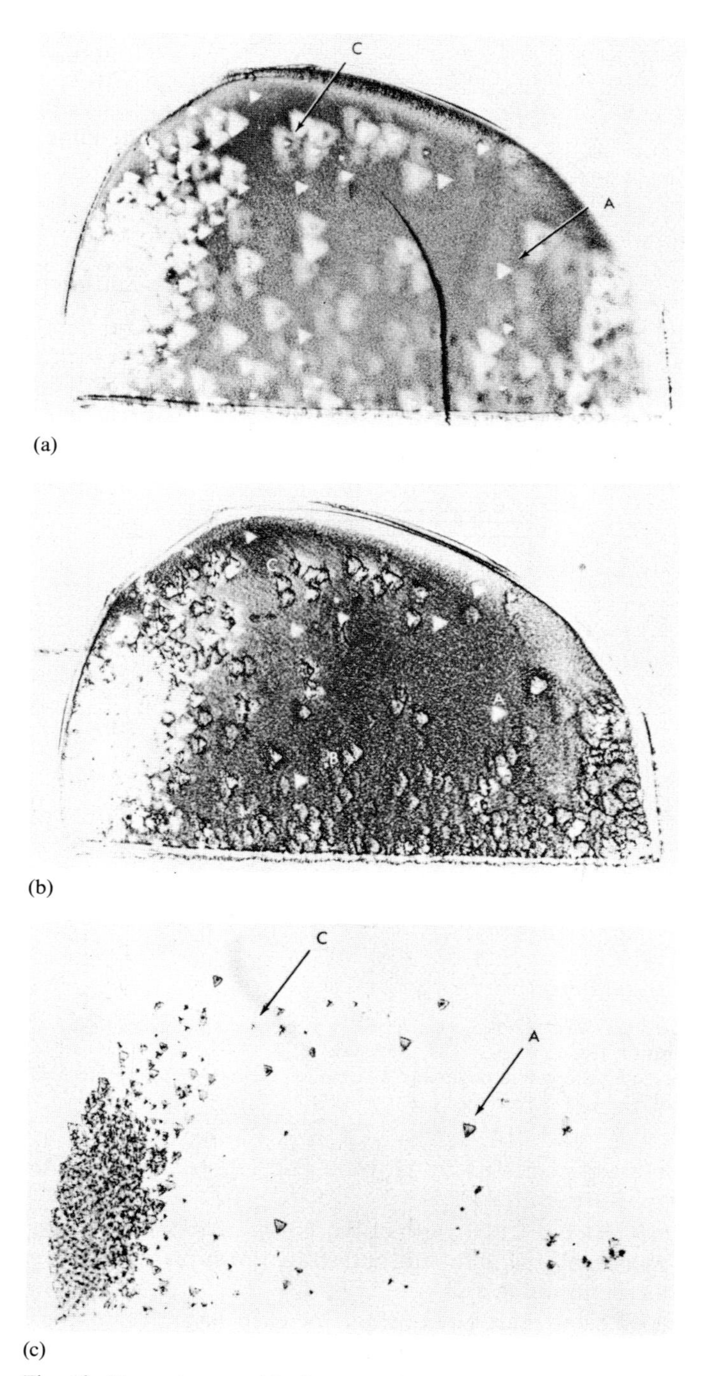


Fig. 10. X-ray topographic images of regions with compositions: (a) GaAs substrate; (b) $GaAs_{0.67}P_{0.33}$ layer; (c) $GaAs_{0.85}P_{0.15}$ hillocks. (By coùrtesy of J. K. Howard.)

For very thick layers the same concepts can also be applied to transmission topographs with the layers on the X-ray exit surface of the sample. Discrimination between peaks is always necessary, if only to separate the $K\alpha_1$–$K\alpha_2$ doublet, and is readily obtained using a device to control the Bragg angle automatically, as described in Section 3.2.2. Another option for better peak discrimination is to narrow the width of the peaks by using radiation that has been monochromatized in a double-crystal spectrometer[24, 25]. (See, for example, the epitaxial lattice misfit compensation studies of Nishizawa *et al.*[26])

*3.2.2. Lateral variations*

Compositional fluctuations, such as growth striations and faceting, distributed laterally across an epitaxial layer or substrate can also be revealed by X-ray topography. One source of image contrast is the angular displacement which occurs in the location of the Bragg reflection due to lattice parameter variations. For example, alternating striae produce increased or decreased reflected intensity depending on the exact setting of the crystal and the local value of the Bragg angle. Strains at the boundary of faceted regions, or between the striations, can also produce contrast which will be enhanced or suppressed depending on the magnitude of the local distortion, the value of $\mu t$, the halfwidth of the rocking curve and the dispersion of the incident beam[27]. Except for topographs imaged in the Borrmann mode these compositional contrast effects are usually independent of the direction of the diffraction vector. This means that a black on grey pair of striations will retain the same contrast throughout a full 360° circle.

In practice, a combination of the lattice mismatch and strain contrast effects occurs. This has been demonstrated by Glass *et al.*[28] and by Stacy and Enz[29] who used compositional X-ray topography to examine heteroepitaxial layers of magnetic garnet deposited onto substrates containing both facets and growth striations. Both lateral and vertical changes in lattice parameter exist simultaneously for these samples and produce a complicated set of strain gradients at each boundary. The magnitude of the compositional changes and the variety of strain effects are discussed in relation to the propagation of magnetic bubble domains in the epitaxial films.

The use of a double-crystal spectrometer in compositional studies[24–26, 28] permits a quantitatively accurate comparison of one part of a substrate/epi-system to another. However, for absolute lattice parameter measurements it is necessary to use an independent technique, *i.e.* one separate from the Lang camera, such as the Bond[30, 31] method of lattice parameter determination.

## 3.3. Cleavage face topography

Another approach to obtaining an in-depth profile of wafer defect distributions is to obtain a reflection topograph from a cleaved cross section, hereafter called CF–XRT. This technique was first described by Schiller[32] and has been used extensively[33–35] to analyze a variety of substrate processing and layer growth phenomena in group III–V compounds. In some geometric respects it is similar to the depth analysis provided by the transmission section topographs already discussed and shown in Fig. 3(b). However, the CF–XRT approach is faster, more generally applicable to any material with a smooth cleavage face, and offers superior contrast at free surface boundaries. Also experimental constraints such as the value of $\mu t$ can be a limiting factor with section topographs.

Experimentally the only difference between setting up a CF–XRT and a normal reflection topograph is that a long thin edge of the sample rather than the larger planar surface is oriented on the topographic camera. Otherwise, the same tables[22] and the geometry of Fig. 6 can be used. For homoepitaxial layers the technique will yield, in a single exposure, a variety of information for various regions of the crystal:

(1) the distribution and density of defects in the bulk and in the epitaxial layers;

(2) a profile of the strain field at the free surface; and

(3) the imperfection density and mismatch strain, if any, at the film–substrate interface.

For heteroepitaxial samples several exposures may be necessary as in the compositional XRT approach. However, interfacial strain gradient effects dominate and are, in fact, the most useful application of CF–XRT to heterojunction studies, as will be shown below.

A CF–XRT study of ohmic contacting procedures to a two-layer (double-tipped liquid phase epitaxy) GaP p–n junction light-emitting diode is shown in Fig. 11. The sample was examined just after the layer was grown, after alloying of an Au ohmic contact, and finally after etching to remove some of the damage introduced during the contacting procedure. The arrows $I_{n/n}$ and $I_{p/n}$ in Fig. 11 locate the interfaces between the n-type substrate and the LPE layer and between the p- and n-type layers, respectively. In the as-deposited condition the interface between the two layers cannot be distinguished except for a single interfacial dislocation (arrow D in Fig. 11(a)). Both layers have a diffracted X-ray intensity and a line defect density lower than that of the substrate. The improved structure of the LPE layer is consistent with chemical etch-pit and electroluminescence studies[36] and explains, in part, why double tipping is employed to place a buffer layer between the active p–n junction interface and the substrate.

After high temperature alloying of an evaporated Au dot, extensive strain fields are created which penetrate up to 40 μm into the layer (Fig. 11(b)). Fortunately, the p-type layer is more than 50 μm thick and lies outside the grossly strained volume under the contact. However, it should be pointed out that the layer–layer interface is delineated in the vicinity of the alloying strain fields (see arrows $I_{p/n}$) whereas between the contacts and in the as-deposited condition the interfacial strains at $I_{p/n}$ are not visible. Another problem introduced during contacting is a uniform surface strain extending to a depth of 5–10 μm, which is due to a very shallow surface "skin" effect. The strain can be easily eliminated by employing a non-preferential etch to remove about 1 μm of surface material (Fig. 11(c)). However, the large strain fields under the contacts are only partially reduced by the etch.

In addition to the strains imaged on as-deposited epi-surfaces, the strain sensitivity of the CF–XRT technique is especially well suited to the detection of residual polishing damage[35] and to the optimization of shaping processes

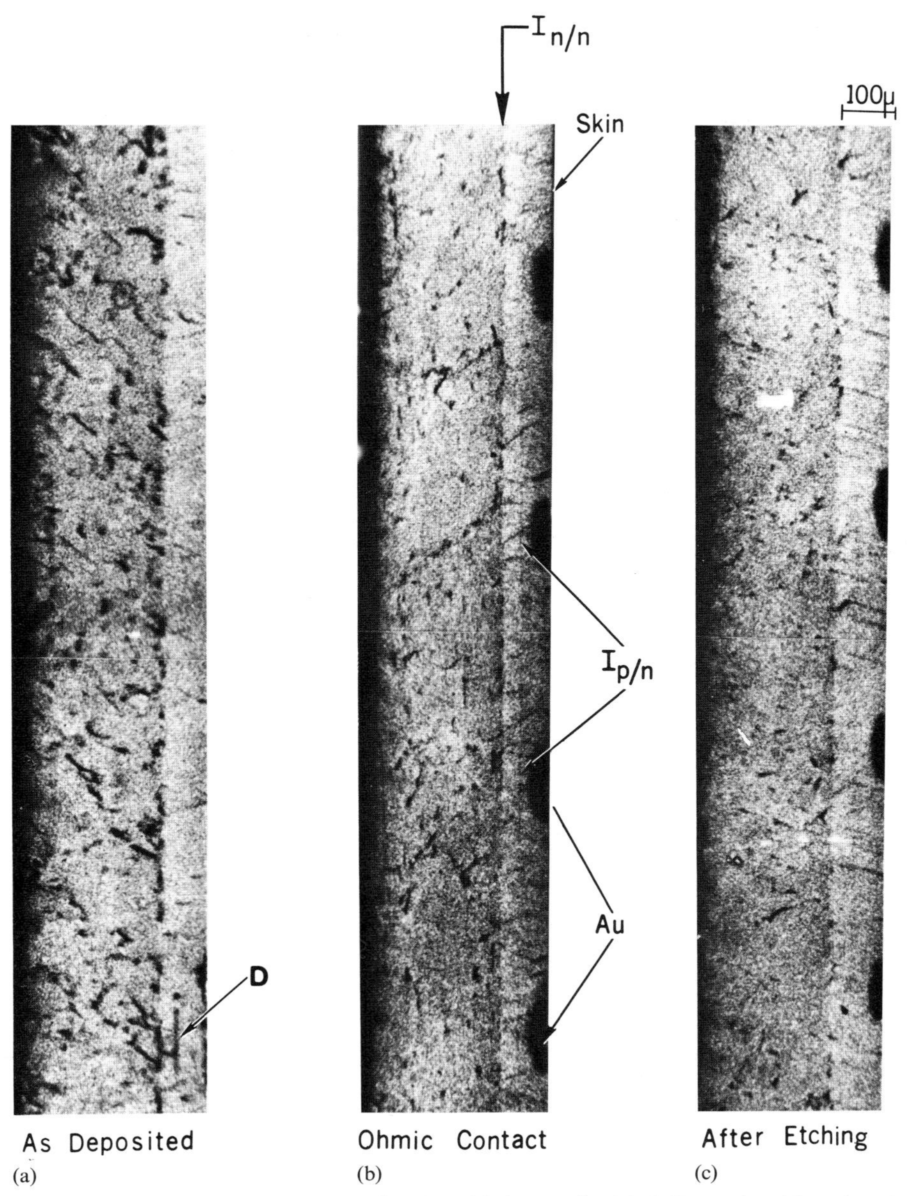


Fig. 11. Cleavage face X-ray topographs of double homoepitaxial GaP samples: (a) as-deposited; (b) after ohmic contacting; and (c) after etching.

for obtaining strain-free substrate surfaces. CF–XRTs were used in this context[37] in evaluating a new polishing solution for the A, or Ga-rich, (111) surface of GaAs, which is not adequately polished by the standard bromine–methanol solution.

With regard to misfit strains due to lattice parameter changes from substrate to layer, we present in Fig. 12 three CF–XRTs from samples representative of the $Ga_{1-x}Al_xAs/GaAs$ epitaxial system[38] with $x = 0, 0.02$ and $0.12$. In Fig. 12(a) the X-ray contrast from the (110) cleavage face of a homojunction, *i.e.* $x = 0$, is uniformly grey from the bulk of the substrate out to the surface of the LPE layer, which is located at the right-hand edge of the CF–XRT. If we add Al so that it is just 2% of the group III elements, *i.e.* $x = 0.02$, then the X-ray intensity is increased due to the strain gradient introduced at the interface and a heavy black line results (Fig. 12(b)). Note that the extent of the strain field is approximately 15 μm, which is six times larger than the layer thickness. The lines CS are cleavage steps and the arrow E points to the edge of the layer below which the CF–XRT shows no strain contrast. The strain sensitivity of the XRT image can be more fully appreciated if we consider the amount of lattice mismatch present in a $Ga_{0.98}Al_{0.02}As/GaAs$ junction. Using the data compiled and analyzed in ref. 39 the room temperature mismatch is about 0.008 Å for $x = 1.0$. For the $x = 0.02$ sample of Fig. 12(b) the mismatch is only 0.00016 Å, which

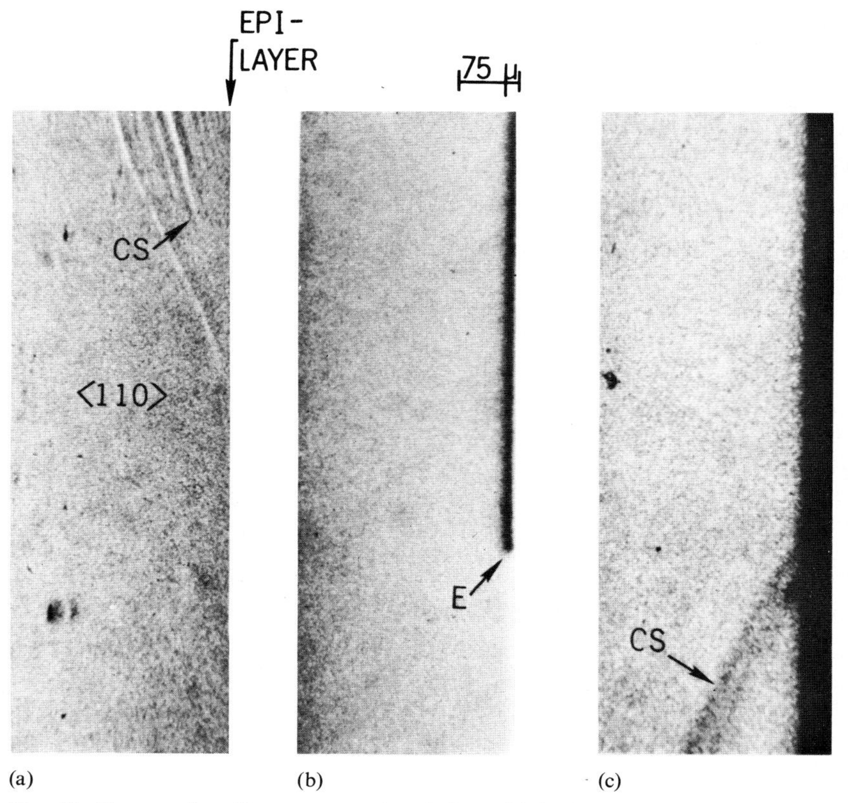


Fig. 12. Cleavage face X-ray topographs of $Ga_{1-x}Al_xAs/GaAs$ heteroepitaxial samples with (a) $x = 0$, (b) $x = 0.02$ and (c) $x = 0.12$. CS are cleavage steps and E is the edge of the epi-layer region on the substrate surface.

is equivalent to a $\Delta a/a$ of $2.8 \times 10^{-5}$. For $x = 0.12$ – an Al concentration about one-half that used for stripe geometry lasers – the strain field extends into the substrate more than an order of magnitude deeper than the 3 μm layer (Fig. 12(c)).

## 4. QUANTITATIVE MEASUREMENTS AND AUTOMATIC CONTROLS

### 4.1. *Substrate curvature*

A constant problem with device material, in particular with silicon integrated circuit wafers, is the warpage introduced during processing, which leads to serious difficulties with the alignment, resolution and wear of photolithographic masks. It is also indicative of the presence of elastic stress gradients and plastic flow discontinuities, such as dislocations or impurity precipitation. It is therefore essential that the magnitude and, if possible, the origin of the warpage be specified. Whether it takes the form of lattice curvature, lattice dilatation or both, warpage leads to X-ray topographs which are not uniformly exposed since the Bragg equation is not uniquely satisfied at all points of the crystal. This presents a troublesome but not insurmountable problem for using X-ray topography as a process control tool.

Because of the known separation of the characteristic $K\alpha_1$ and $K\alpha_2$ peaks it is possible to determine the wafer radius of curvature very quickly and accurately. This is illustrated in Fig. 13 which is a (422) reflection topograph recorded on dental X-ray film after a 12 min exposure. The two images occur because on the right the Bragg equation is satisfied for the $K\alpha_1$ peak while on the left the $K\alpha_2$ peak gives an image which is half the intensity of the $K\alpha_1$ peak. The simple calculation required for determining the radius of curvature is shown in the figure.

The double-image technique has been used to solve a warpage and cracking problem with GaAs IMPATT devices[40] which use electroplated Au or Ag heat sinks. Basically, a multilayer metallization scheme is employed using layers stressed either in tension or in compression so that the summation of all curvatures leads to stress compensation in the GaAs device, *i.e.* a flat wafer.

If the radius of curvature is much larger than 10 m for a wafer 2–3 cm wide only one image will be recorded on the topograph. Also, if the substrate is warped because of the presence of growth faults the images may overlap or even be indistinguishable. This is shown in Fig. 14(a) for a grossly dislocated silicon wafer which contains regions slightly misoriented with respect to one another.

Samples of this type can nevertheless be analyzed quite thoroughly if an automatic Bragg angle control (called ABAC) device, described below, is used to maintain the orientation of the sample so that the $K\alpha_1$ Bragg peak is continuously recorded. Using an ABAC during the exposure yields the data shown in Fig. 14(b). Note that the X-ray image is much more uniformly exposed, except in the four corners of the (100) wafer where the dislocation density was too high to permit any imaging in the Borrmann mode. The magnitude of the lattice tilt across the two large tilt boundaries, 60″ and 140″, can be determined from the ABAC trace which is superimposed over the topograph. The overall

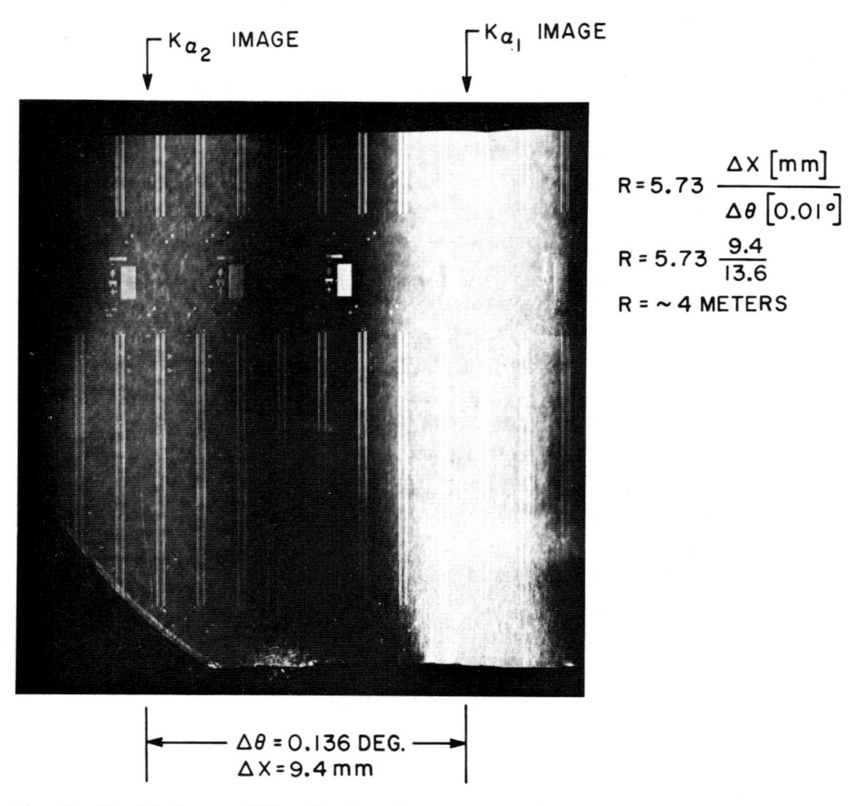


Fig. 13. Double image (422) reflection X-ray topograph of a warped silicon device wafer with a 4 m radius of curvature.

radius of curvature, which is inversely proportional to the average slope of the ABAC trace, *i.e.* $\Delta\theta_B$ over $\Delta X$, is found to be about 20 m. The equipment needed to obtain ABAC topographs and traces, as well as its application to the quantitative determination of stress, will be discussed in the next section.

### 4.2. *Automatic Bragg angle control*

The essential components of an ABAC system[41] are illustrated in Fig. 15. A commercial Lang XRT camera which already has a $\Delta X$ scanning motor has had a stepping motor added to control changes in $\theta_B$ through rotation of the camera base. Ten-turn potentiometers are used to couple an X–Y recorder to the $\Delta X$ and $\Delta\theta_B$ motions, and the diffracted X-ray intensity can also be fed to the recorder. The inputs to the X–Y recorder can be selected so that automatic plotting of any of the following can be obtained:

(1) rocking curves – intensity *versus* Bragg angle;
(2) ABAC traces – change in Bragg angle *versus* distance; and
(3) sample uniformity – intensity *versus* distance across the sample.

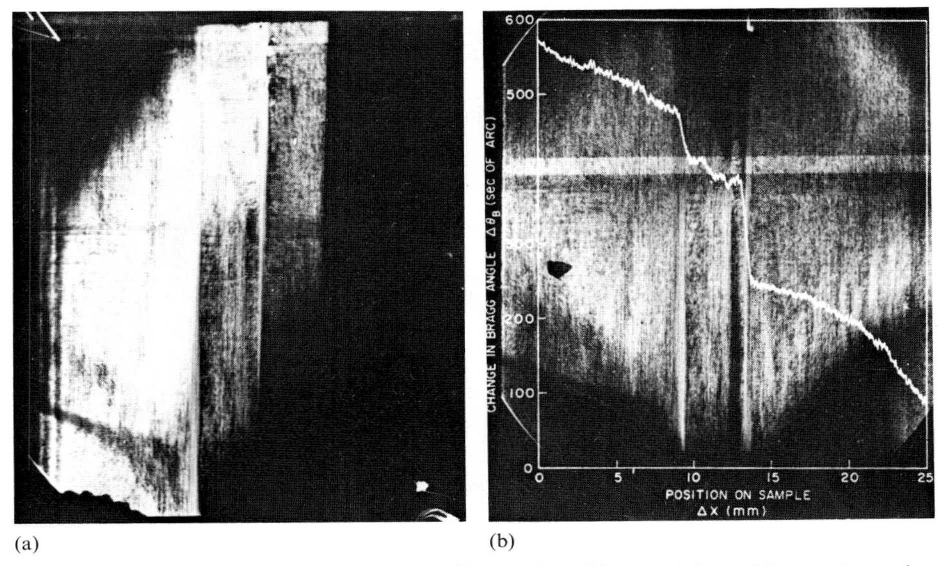


(a) (b)

Fig. 14. Transmission X-ray topographs of a silicon wafer with gross defects: (a) normal scanning mode; (b) ABAC with superimposed $\Delta\theta$ *vs.* $\Delta X$ trace.

Details of the integrated circuit and associated electronics that make up the most recent ABAC feedback electronics have been described by Storm[42]. Basically, the unit functions as an operational amplifier whose output signal is positive or negative depending on whether the X-ray diffracted beam intensity, *i.e.* output from the Geiger tube ratemeter, exceeds or falls below predetermined upper and lower limits which we call the ABAC window. The ABAC output is then used to drive the stepping motor which rotates the crystal in the appropriate direction to keep the X-ray beam intensity within the ABAC window.

The ABAC window can best be visualized by referring to Fig. 16, which is an actual (220) rocking curve obtained from a fixed geometric position on a (111) silicon wafer. This curve was obtained by automatically plotting the change in intensity for each incremental change in wafer orientation in the vicinity of the K$\alpha$ doublet. Recall the two images in the warped crystal of Fig. 13. The operational amplifier, which is the main component of the ABAC electronics, is then set to operate in a window defined by $I_{80}$ and $I_{70}$. If the intensity exceeds $I_{80}$ the stepping motor is activated so as to change the Bragg angle in a positive direction, while a decrease to $I_{70}$ steps the motor in the opposite direction. If a warped crystal is set in motion by activation of the X-scan on the Lang camera, the ABAC electronics and stepping motor will maintain the orientation of the wafer to within about 6″ while the X–Y recorder plots each change in $\theta_B$ as a function of sample position. The 6″ is the amount of rotation imparted to the table of the Lang camera for a window operating between 70 and 80% of $I_{max}$.

van Mellaert and Schwuttke[43] have built a considerably more sophisticated, and expensive, ABAC system which maintains the wafer orientation at $I_{max}$,

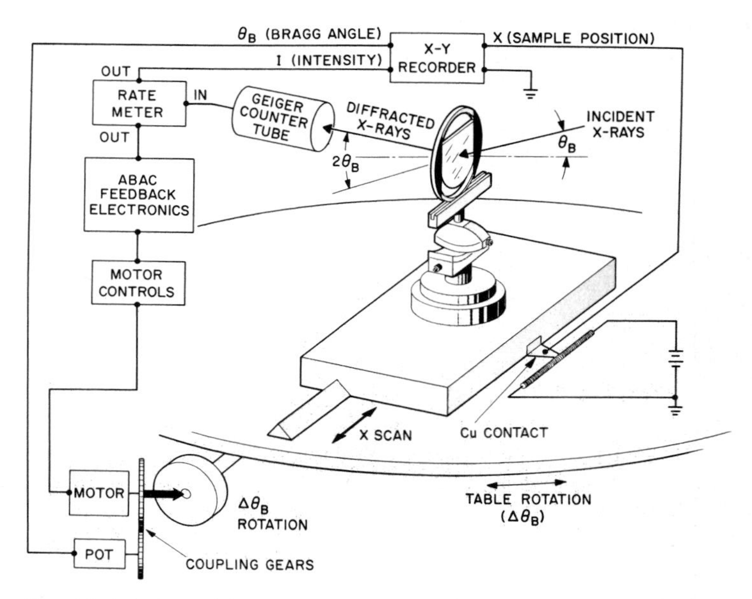


Fig. 15. Schematic diagram of the automatic Bragg angle control (ABAC) equipment.

yielding higher overall intensity and shorter XRT exposures. Also, their system will accommodate large excursions in the diffracted beam intensity and still ride the $I_{max}$ peak, whereas riding the shoulder of the rocking curve may lead to difficulties if $I_{max}$ is not reasonably constant. Although van Mellaert and Schwuttke did not describe an extension of their ABAC to the actual plotting of $\Delta\theta$ *versus* $\Delta X$ it could readily be achieved. In any case, the ABAC topographs are a major improvement over the scanning oscillator technique (SOT) previously used[44] on warped wafers, since only the $K\alpha_1$ peak is imaged and no time is wasted in oscillating off the peak. ABAC would also be particularly useful with the narrow $K\alpha_1$ peaks obtained in a double-crystal diffractometer. However, for diffraction from one crystal the best sensitivity for curvature, and ultimately layer stress, is obtained by riding the steep shoulder of the rocking curve shown in Fig. 16 rather than the broader peak, as shown in the next section.

*4.3. Stress calculations and sensitivity*

Once the ABAC trace is obtained the radius of curvature can be calculated from the simple formula

$$R = 206\ \Delta X/\Delta\theta_B \qquad (1)$$

where $X$ is in millimeters, $\Delta\theta_B$ is in seconds of arc and $R$ is in meters. By monitoring the steep $I_{70}$ to $I_{80}$ window across a 30 mm sample, which yields a 6″ change in orientation, radii of curvature as large as 1000 m can be determined. Riding $I_{max}$ down to $I_{90}$ across the broader peak would be adequate for radii of 150 m or less.

Once the radius of curvature is known the stress in the thin film[45] can be calculated using the equation

$$\sigma_f = \frac{E}{6(1-\nu)} \frac{t_s^2}{t_f} \frac{1}{R} \tag{2}$$

where $t_s$ is the substrate thickness, $t_f$ is the film thickness, and $E$ and $\nu$ are Young's modulus and Poisson's ratio for the substrate. For (111)-oriented silicon wafers with $t_s = 250$ μm eqn. (2) can be expressed as

$$\sigma_f = \frac{248 \times 10^9}{R\, t_f} \quad \text{dyn} \quad \text{cm}^{-2} \tag{3}$$

where $R$ is in meters and $t$ in $\text{Å} \times 10^3$. Therefore the minimum stress that can be determined is inversely related to the product of the film thickness and the maximum radius of curvature that can be accurately measured. This product would be $25 \times 10^{10}$ dyn cm$^{-2}$ Å$^{-1}$ for an ABAC unit with a 6″ sensitivity, *i.e.* $R_{max} \sim 1000$ m and substrates 250 μm thick. For example, minimum stresses of $2.5 \times 10^8$ or $2.5 \times 10^7$ dyn cm$^{-2}$ are obtained for 1000 Å or 10 000 Å films. For yet more sensitive work, thinner substrates and double-crystal narrowing of the K$\alpha_1$ peak can be used.

Once the film stress is calculated, a value for the maximum stress in the substrate can be obtained using the equation

$$\sigma_{s,\,max} = 4\,(t_f/t_s)\,\sigma_f \tag{4}$$

An estimate of $\sigma_{s,\,max}$ for a 2500 Å tungsten film with $\sigma_f = 2.5 \times 10^{10}$ dyn cm$^{-2}$ on a Si wafer 250 μm thick yields a value of $1 \times 10^8$ dyn cm$^{-2}$. This is considerably lower than the threshold stresses required for dislocation motion in Si at high temperatures. Therefore no plastic flow phenomena are expected in the substrate due to the presence of a highly stressed thin film. However, it must be pointed out that for actual device configurations the value of stress at sharp corners and step edges can be significantly higher, as indicated in Fig. 1(a), the schematic diagram in Fig. 4 and in Section 4.4.

### *4.4. ABAC applications: multilayer device*

The ABAC system has been used to plot local changes in substrate curvature in a scaled-up version of a multilayer metal–oxide–semiconductor device structure. The layer profiles in actual use were a 0.1 μm gate oxide and a 0.7 μm field oxide, each of which was covered with 0.2 μm of tungsten. A sample was generated for ABAC analysis by cutting 5 mm wide stripes in a 0.7 μm steam oxide, re-oxidizing the entire wafer to grow 0.1 μm of dry oxide, and then sputter-depositing 0.2 μm of tungsten over the entire wafer. A set of 5 mm stripes were etched in the tungsten so that half of each stripe was over thick or thin oxide, leaving $2\frac{1}{2}$ mm of thick and thin oxide uncovered. The inset to Fig. 17 gives a plan view of the final structure, while the ABAC trace A represents the local substrate

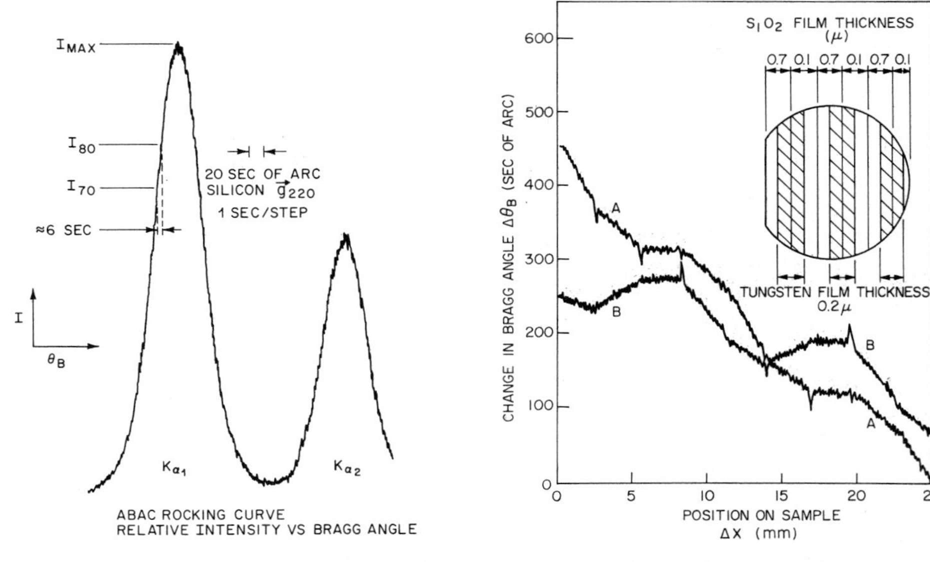


Fig. 16. Rocking curve obtained on ABAC equipment illustrating the operating window and sensitivity.

Fig. 17. ABAC analysis of a $W/SiO_2/Si$ sample: (a) before annealing; (b) after a 900 °C annealing.

curvature for the as-deposited sample and trace **B** is from the same sample after a 900 °C annealing. Each $2\frac{1}{2}$ mm segment of the ABAC trace yields information on a different layer or multilayer stress situation and is repeated every fourth segment. The layer sequence is as follows: starting from the left, W on thick oxide, W on thin oxide, thin oxide, thick oxide, W on thick oxide etc. Since the as-deposited W and $SiO_2$ are both compressively stressed the substrate curvature, *i.e.* the slope of the ABAC trace, is negative throughout trace A, with the steepest slope corresponding to a W on thick oxide multilayer. After annealing at 900 °C the sign and magnitude of the W stress changes from highly compressive to moderately tensile, whereas the oxide stresses are unchanged. As a result, the W on thin oxide segment has a positive slope in trace **B**, while the W on thick oxide segment is reduced in slope but remains negative. This is due to the combined effects of the layer thickness (three times larger for the oxide) and the magnitude of stress (about two times larger for W). Although the $2\frac{1}{2}$ mm wide stripes are by no means of semiconductor device dimensions, we are at least in a position to obtain quantitative data on the stresses (both for the layer and for the substrate) introduced during each step in the processing of various structures.

It must be pointed out that, although eqn. (4) indicates that the substrate stresses will be up to 100 times less than the layer stress, the substrate stress at sharp corners can be significantly higher (note the discontinuities in the ABAC traces at the edges of the stripes in Fig. 17). Unfortunately, the quantitative aspects of the ABAC technique break down at regions of non-uniform X-ray intensity and the actual stress increase can only be roughly estimated, as discussed in relation to Fig. 1 and in Section 4.2.

Figure 1 also indicates that the generation of dislocations and the formation of intermetallic compounds can be a major factor in high temperature processes. Therefore, for an accurate stress determination in these cases a topographic survey of possible plastic flow phenomena is essential to ensure that none of the elastic energy available for lattice distortion has been released in generating dislocations. A combined ABAC–XRT approach is particularly important where threshold strains required for misfit dislocation formation in heteroepitaxial systems are of interest[39].

*4.5. High temperature ABAC*

Ideally the MOS annealing experiments described in the preceding section would be done *in situ* on a hot stage so that both the ABAC traces and the X-ray topographs would be obtained while the sample was at a high temperature. Although a hot stage was not available at the time of the MOS studies, subsequent work on GaAlAs/GaAs heteroepitaxial layers was performed at sample temperatures up to 350 °C. The GaAlAs/GaAs system is especially well suited to illustrate the use of what we call the "hot ABAC" technique because it has been shown[47] that the lattice parameters of these compounds (powdered AlAs and single-crystal GaAs) are matched at about 900 °C but mismatched at room temperature. Therefore, if the epitaxial lattice parameters are similar to the bulk values, then the curvature of the epi-layer/substrate combination should decrease with increasing temperature and approach zero near 900 °C. This reduction of lattice misfit with temperature has been verified, as shown by the hot ABAC traces in Fig. 18.

Taking a linear projection of these curvature data out to the growth temperature yields an intercept for zero curvature, *i.e.* perfect lattice matching, at 830 °C, as shown in Fig. 19. This compares very favorably with the non-epitaxial perfect match estimate and is certainly within the experimental error of the bulk lattice parameter and epi-ABAC measuring techniques. For example, the ABAC traces in Fig. 18 actually have two regions, right and left, with different slopes. This non-uniformity in the data can be traced to a left-to-right variation in the thickness of the epi-layer. It is necessary in these cases to cleave the sample where the ABAC trace was made and obtain a thickness profile. Substituting local values of $R$ and $t_f$ in eqn. (2) yields a layer stress that is constant across the sample for a given temperature. An X-ray topograph of the entire wafer with an extra scan where the ABAC trace was taken, such as is shown in Fig. 14(b), pinpoints the location of the cleavage line.

If the bulk properties of the substrate are known as a function of temperature it is then possible to obtain absolute values of lattice parameter, thermal expansion coefficient and certain elastic constants of the film using the hot ABAC technique. Also, for situations where plastic flow is initiated, the ratio of elastic energy accommodated by the lattice and that absorbed by the dislocations can be deduced, since a complementary XRT would image the type and density of the dislocations introduced.

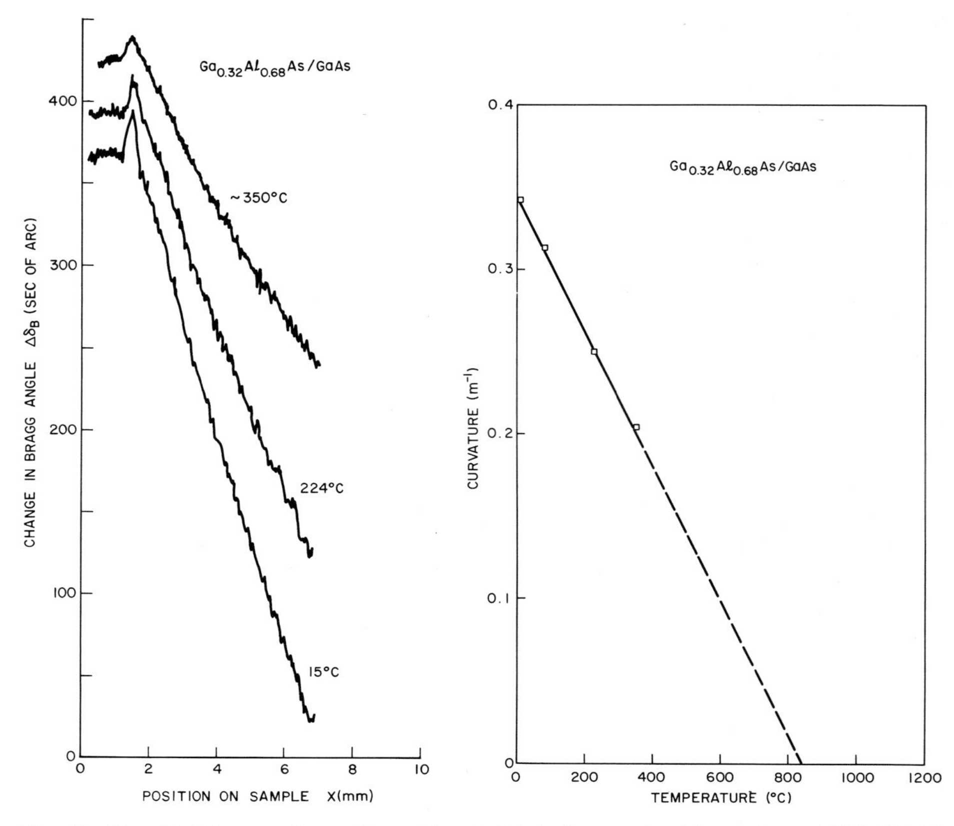


Fig. 18. Hot ABAC traces for a $Ga_{0.32}Al_{0.68}As/GaAs$ heteroepitaxial sample at 10 °C, 224 °C and 350 °C.

Fig. 19. Heterojunction curvature as a function of temperature for the data of Fig. 18.

## 5. COMBINED X-RAY APPROACH

This paper has presented a number of analytical techniques which can all be performed on the same instrument—a Lang X-ray camera. The various examples were chosen to illustrate best the particular analysis or mode of operation under discussion. However, because of their complementary nature, the real power in the application of these tools lies in their combined use. As a final example, we describe how each of the experimental techniques described can be applied in a complementary way to a single technological problem—the fabrication of a heterojunction laser of GaAs sandwiched between layers of $Ga_{1-x}Al_xAs$ which are lattice-matched at room temperature and free of dislocations. Since these two factors, stress[48] and dislocations[49], have been demonstrated to limit the life of lasers, we have systematically measured the stress, the lattice parameter differences, the dislocation content, and, where possible, the high temperature behavior of the heteroepi-system. A typical sequence in which the X-ray techniques were used is:

(1) transmission X-ray topography;
(2) ABAC;
(3) rocking curve analysis;
(4) compositional X-ray topography;
(5) cleavage face X-ray topography;
(6) hot ABAC.

Transmission X-ray topography showed that the substrate was only contributing from $10^2$ to at most $10^4$ dislocations per square centimeter to the epitaxial layer. No new dislocations were generated during the epitaxial growth. The hot ABAC data presented in Figs. 18 and 19 confirmed that lattice matching at the growth temperature allowed a completely elastic layer/substrate situation to prevail. The amount of elastic stress was then quantitatively determined as a function of Al content using the ABAC device[38], while cleavage face topographs (Fig. 12) gave images of the heterojunction strain.

At this point it was realized that stress compensation at room temperature could be achieved by adding small amounts of GaP to the GaAlAs melt in order to match the lattice parameters of the waveguiding layers to the active GaAs layer[50]. The amount of P required was predicted from Vegard's law considerations to be about 1.5% P for an Al content of 34%. This was experimentally verified as shown by the rocking curve peak shifts, *i.e.* compositional analysis, shown in Fig. 20. Note that in the trace A the layer K$\alpha$ peaks are to the right of the substrate reference, indicating a larger lattice parameter and compressive stress in the layer. Adding about 4% P shifts the layer peaks to the left, as shown

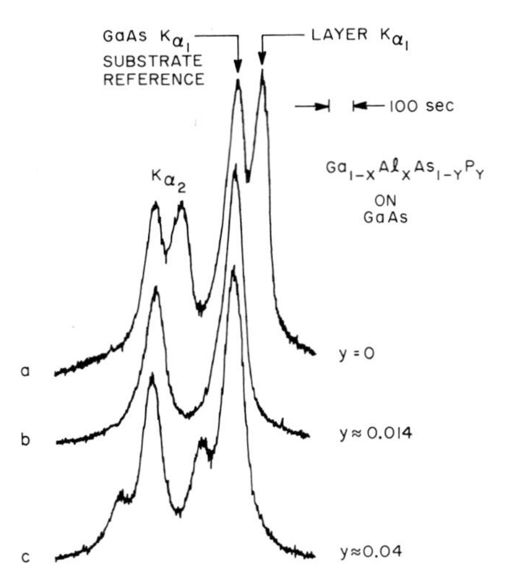


Fig. 20. Heterojunction rocking curve analysis as a function of P content.

by trace C, producing a layer with a tensile stress. Stress compensation occurred for $y \approx 0.015$ (trace B), for which only two peaks are recorded in the rocking curve since the K$\alpha$ doublet of the layer is superimposed on that of the substrate.

Once the P was added, however, transmission X-ray topographs of certain GaAlAsP samples (see, for example, Fig. 5) showed that misfit dislocations were generated due to the lattice mismatch which now occurred at the growth temperature. It was also observed that the length of the misfit dislocation could be extended to the edge of the sample (see defects A and B in Fig. 5(a)). In this way a means for defect elimination from the active laser epi-layer by the deliberate introduction of misfit dislocations was achieved[51]. Careful control of the epi-layer thickness for P concentrations corresponding to $y \approx 0.015$ created room temperature stress compensation (a stress-free layer), while the stresses which were present at the growth temperature served to create misfit dislocations which swept out substrate dislocations (a dislocation-free layer)[39].

Additional X-ray diagnostic work can be extremely useful in maintaining a low stress defect-free sample throughout the many processes required in the fabrication of an actual laser. For example, if new defects are introduced due to shaping and contacting procedures, they can be monitored and the appropriate processing steps re-evaluated using the curvature and defect imaging techniques described above.

## 6. CONCLUSION

In this report we have attempted to describe how a single Lang X-ray camera can make a broad range of diagnostic techniques available to a materials scientist involved in thin film electronics research or development technology. The principles of image formation for transmission and reflection X-ray topography have been briefly reviewed with both plastic (dislocations) and elastic (layer stress) phenomena illustrated. The procedures required in the analysis of different types of dislocations as well as in the quantitative determination of layer and substrate stresses have also been presented, along with previously unpublished high temperature stress measurements of a GaAlAs heteroepitaxial layer. It has been shown that selected use of reflection or transmission geometries provides imaging of individual epi-layers or entire multilayer samples, depending on the information desired. Many other materials analysis problems related to electronics technology, *e.g.* both grown-in or process-induced defects, as well as the determination of the properties of complex dielectric, metallic or epitaxial structures, can be readily formulated by the combined application of the techniques described above.

A final comment on the use of a combined X-ray approach is that it should only be considered in the light of the problem at hand. There are times when simple optical microscopy of chemical etch pits, or the displacement of interference fringes, provides adequate information. On other occasions, transmission electron microscopy or Auger analysis is required to provide a combined approach which is larger in scope than the X-ray studies alone and is, in fact, one of the objectives of this special issue on characterization techniques.

ACKNOWLEDGMENTS

The authors would like to thank collectively the various materials scientists who supplied the samples described in this report, as well as J. K. Howard and S. H. McFarlane who provided Figs. 8, 9, 10 and Fig. 3, respectively. Comments on the final manuscript by I. A. Blech, R. B. Marcus and J. W. Nielsen are also appreciated.

REFERENCES

1 A. R. Lang, *J. Appl. Phys.*, *5* (1958) 358.
2 A. R. Lang, *Acta Crystallogr.*, *12* (1959) 249.
3 A. R. Lang, in S. Amelinckx (ed.) *Modern Diffraction and Imaging Techniques in Materials Science*, North Holland Publ. Co., Amsterdam, 1970, pp. 407–480.
4 A. Authier, in S. Amelinckx (ed.) *Modern Diffraction and Imaging Techniques in Materials Science*, North Holland Publ. Co., Amsterdam, 1970, pp. 481–520.
5 E. S. Meieran, *Siemens Rev. 4th Special Issue*, *37* (1970) 1–36.
6 G. H. Schwuttke, *Microelectron. Reliab.*, *9* (1970) 397–412.
G. H. Schwuttke, in J. W. Matthews (ed.), *Epitaxial Growth*, Part A, Academic Press, New York, 1975.
7 H. Kressel, P. Robinson, R. V. d'Aiello and S. H. McFarlane, *J. Appl. Phys.*, *45* (1974) 3930.
8 G. Borrmann, *Z. Phys.*, *127* (1950) 297.
9 P. Penning and D. Polder, *Philips Res. Rep.*, *16* (1961) 419.
10 E. S. Meieran and I. A. Blech, *Phys. Status Solidi*, *29* (1968) 653.
11 G. H. Schwuttke and J. K. Howard, *J. Appl. Phys.*, *39* (1968) 1581.
12 Y. Ando, J. R. Patel and N. Kato, *J. Appl. Phys.*, *44* (1973) 4405.
13 E. S. Meieran and I. A. Blech, *J. Appl. Phys.*, *43* (1972) 265.
14 S. Mader, A. E. Blakeslee and J. Angilello, *J. Appl. Phys.*, *45* (1974) 4730.
15 S. Kishino and M. Ogirima, *Philos. Mag.*, *31* (1975) 1239.
16 W. F. Berg, *Naturwissenschaften*, *19* (1931) 391.
17 C. S. Barrett, *Trans. Metall. Soc. AIME*, *161* (1945) 15.
18 J. B. Newkirk, *Trans. Metall. Soc. AIME*, *215* (1959) 483.
19 A. P. L. Turner, T. Vreeland, Jr. and D. P. Pope, *Acta Crystallogr.*, *A24* (1968) 452.
20 R. W. James, *The Optical Principles of the Diffraction of X-Rays*, Cornell Univ. Press, Ithaca, N.Y., 1965.
21 R. C. Blish, II, *Adv. X-Ray Anal.*, *14* (1971) 163.
22 M. A. G. Halliwell, J. B. Childs and S. O'Hara, *Proc. 1972 Symp. on GaAs*, Inst. Phys., London, 1973, pp. 98–105.
23 J. K. Howard and R. D. Dobrott, *J. Electrochem. Soc.*, *113* (1966) 567.
24 B. G. Cohen, *Solid-State Electron.*, *10* (1967) 33.
25 M. Hart and K. H. Lloyd, *J. Appl. Crystallogr.*, *8* (1975) 42.
26 Y. T. Lee, N. Miyamoto and J. Nishizawa, *J. Electrochem. Soc.*, *122* (1975) 530.
J. Nishizawa, T. Terasaki, K. Yagi and N. Miyamoto, *J. Electrochem. Soc.*, *122* (1975) 664.
27 D. C. Miller and A. F. Witt, *J. Cryst. Growth*, *29* (1975) 19.
28 H. L. Glass, P. J. Besser, T. N. Hamilton and R. L. Stermer, *Mater. Res. Bull.*, *8* (1973) 309.
29 W. T. Stacy and U. Enz, *IEEE Trans. Magn.*, *8* (1972) 268.
30 W. Bond, *Acta Crystallogr.*, *13* (1960) 814.
31 R. L. Barns, *Adv. X-Ray Anal.*, *15* (1972) 330.
32 C. Schiller, *J. Appl. Crystallogr.*, *2* (1969) 223.
33 C. Schiller, *Solid-State Electron.*, *13* (1970) 1163.
34 G. A. Rozgonyi and S. E. Haszko, *J. Electrochem. Soc.*, *117* (1970) 1562.
35 G. A. Rozgonyi and R. H. Saul, *J. Appl. Phys.*, *43* (1972) 1186.
36 G. A. Rozgonyi and T. Iizuka, *J. Electrochem. Soc.*, *120* (1973) 673.
37 J. C. Dyment and G. A. Rozgonyi, *J. Electrochem. Soc.*, *118* (1971) 1346.

38 G. A. Rozgonyi, C. J. Hwang and T. J. Ciesielka, *J. Electrochem. Soc.*, *120* (1973) 333C.
39 G. A. Rozgonyi, P. M. Petroff and M. B. Panish, *J. Cryst. Growth*, *27* (1974) 106.
40 G. A. Rozgonyi, J. V. DiLorenzo and E. Heinlein, *J. Electrochem. Soc.*, *121* (1974) 426.
41 G. A. Rozgonyi and T. J. Ciesielka, *Rev. Sci. Instrum.*, *44* (1973) 1053.
42 A. R. Storm, *Rev. Sci. Instrum.*, *46* (1975) 883.
43 L. van Mellaert and G. H. Schwuttke, *J. Appl. Phys.*, *43* (1972) 687.
44 G. H. Schwuttke, *J. Appl. Phys.*, *36* (1965) 2712.
45 A. Brenner and S. Senderhoff, *J. Res. Natl. Bur. Stand.*, *42* (1949) 105.
46 R. J. Jaccodine and W. A. Schlegal, *J. Appl. Phys.*, *37* (1966) 2429.
47 M. Ettenberg and F. J. Paff, *J. Appl. Phys.*, *41* (1970) 3926.
48 R. L. Hartman and A. R. Hartman, *Appl. Phys. Lett.*, *23* (1973) 147.
49 P. M. Petroff and R. L. Hartman, *Appl. Phys. Lett.*, *23* (1973) 469.
50 G. A. Rozgonyi and M. B. Panish, *Appl. Phys. Lett.*, *23* (1973) 533.
51 G. A. Rozgonyi, P. M. Petroff and M. B. Panish, *Appl. Phys. Lett.*, *24* (1974) 251.